T0360571

SCIENCE IN SOCIETY

Climate Change and Climate Policies

SCIENCE IN SOCIETY

Climate Change and Climate Policies

Nico Stehr
Zeppelin University, Germany
University of Alberta, Canada

Hans von Storch
Helmholtz-Zentrum Hereon, Germany
Hamburg University, Germany

World Scientific

NEW JERSEY • LONDON • SINGAPORE • BEIJING • SHANGHAI • HONG KONG • TAIPEI • CHENNAI • TOKYO

Published by

World Scientific Publishing Europe Ltd.

57 Shelton Street, Covent Garden, London WC2H 9HE

Head office: 5 Toh Tuck Link, Singapore 596224

USA office: 27 Warren Street, Suite 401-402, Hackensack, NJ 07601

Library of Congress Cataloging-in-Publication Data
Names: Stehr, Nico, author. | Storch, H. v. (Hans von), 1949– author.
Title: Science in society : climate change and climate policies /
Nico Stehr, Zeppelin University, Germany, University of Alberta, Canada,
Hans von Storch, Helmholtz-Zentrum Hereon, Germany, Hamburg University, Germany.
Description: New Jersey : World Scientific, [2024] | Includes bibliographical references and index.
Identifiers: LCCN 2022041382 | ISBN 9781800613515 (hardcover) | ISBN 9781800613522 (ebook) |
ISBN 9781800613539 (ebook other)
Subjects: LCSH: Climatic changes--Social aspects. | Climatic changes--Political aspects. |
Climatic changes--Economic aspects. | Science and state.
Classification: LCC QC903 .S737 2024 | DDC 363.738/74--dc23/eng 20221121
LC record available at https://lccn.loc.gov/2022041382

British Library Cataloguing-in-Publication Data
A catalogue record for this book is available from the British Library.

For any available supplementary material, please visit
https://www.worldscientific.com/worldscibooks/10.1142/Q0399#t=suppl

Desk Editors: Balasubramanian Shanmugam/Adam Binnie/Shi Ying Koe

Typeset by Stallion Press
Email: enquiries@stallionpress.com

Printed in Singapore

Preface

When we, Hans von Storch and Nico Stehr, the authors of the contributions to this anthology, set out reflecting about the popular and scientific perception and construction of the phenomenon of climate, climate change, climate policy and the impact of climate on society in the early 1990s, we encountered notable resistance especially as we wrote about the urgent need for societal adaptation to climate change. We see ourselves as pioneers in this policy field. Something is wrong with our planet, and it is obvious that this has to do with human conduct.[1] Many, if not the majority, of countermeasures to climate impacts require the innovative capacity of science and technology. However, the translation of scientific knowledge into society is not an automatic or autonomous force.[2] Moving science into society is subject to economic, political and cultural constraints.

Nico Stehr is a sociologist specializing in the theory of modern society and the sociology of knowledge; Hans von Storch is a mathematician and is also a physical climate scientist.

[1] Consult Dipesh Chakrabarty, *The Climate of History in a Planetary Age*. Chicago: University of Chicago Press, 2021. According to the 2021 European State of the Climate Report (ESOTC) compiled by the Copernicus Climate Change Service (C3S) and implemented by the European Centre for Medium-Range Weather Forecasts (ECMWF) on behalf of the European Commission, the summer of 2021 was the warmest in Europe since records began. The summer months were on average about 1 °C warmer than those of the last three decades.

[2] Roger Pielke Jr., *The Honest Broker: Making Sense of Science in Policy and Politics*. Cambridge: Cambridge University Press, 2007.

Since we "inhabit" rather different scientific cultures, our collaboration is genuinely interdisciplinary[3]: One culture is mainly concerned with social processes and change, and the other with the change and impact of climate on the real world as the natural sciences understand it.

Of course, there have always been reflections in the past on how society and climate and its change are related. For centuries, *climatic determinism* (see Chapter 3 in this volume) dominated the reflections about the relation between society and climate until it finally lost its legitimacy, at least in science, in the mid-20th century, only to be resurrected in a new, modern version — in much of the thinking of environmental economists and theoretical physicists.[4] But some of the intriguing questions about the practice of climate science concern the social process of scientific reasoning itself, for example, why ideas are extraordinarily resistant to change, or how the ways of scientists acting in the societal realm are at best given secondary attention or not examined at all.

Considering the long history of reflection on the interrelationship between climate and society, few core social scientists have addressed the climate issue in recent decades. Perhaps they felt locked out of the natural science approach to the climate issue, while the natural sciences succumbed to the temptation of versions of climate determinism as they reflected about the impact of climate on societies.

In short, both the natural and the social sciences sides relied on the prevailing *Zeitgeist* inside and outside of science rather than engaging with the existing distinctiveness of the other discipline to develop collaborative research efforts. We believe we have successfully initiated such an approach. In our anthology, we outline how we perceive our contribution to the history of ideas in climate science and note that a wide-ranging, almost obvious and necessary collaboration between social scientists and climate scientists has more and more emerged in the most recent decades.

More specifically, our anthology documents the interdisciplinary path and the wide range of themes that has occupied us during more than three decades of joint research and writing and that continue to be of benefit to current research and reflection on the interrelation between nature, democracy, society, governance and climate.

---

[3] Cf. Peter Weingart and Nico Stehr (eds.), *Practising Interdisciplinarity*. Toronto: University of Toronto Press, 2000, for a range of case studies on interdisciplinary research activities.

[4] See Stehr, N.: "In-between: The simultaneity of the non-simultaneous," *Soc. Epistemol.* 36(4): 407–424, 2022.

Some of the articles are about 30 years old, and others are more recent — but we are confident that they have neither lost their analytical depth nor their societal significance.

Finally, a brief remark about the title of our collection: The subtitle is self-explanatory. *Climate Change and Climate Policies* covering the broad range of our inquiries into the dynamics of the interrelation between society and climate and the increasingly conscious intervention of societies into climate. The main title *Science in Society* denotes one of the unique attributes of our research interests: Climate change and the phenomenon of the "Anthropocene" within historical times are a discovery of the sciences rather than a scientific field that has responded to urgent political, economic or health issues. The very fact of climate change comes from the sciences. Our understanding of climate change, for the most part, continues to be inseparable from the knowing and the instruments of climatologists. Hence, science in society. Today, of course, and in the future, societal concerns continue to be very much on the agenda of climate science.

Any errors that we detected in the original publication of the contributions to our anthology have been tacitly corrected. In a few cases, we have updated the literature. Moreover, it is in the nature of such a collection that redundancies cannot be completely avoided.

About the Authors

Nico Stehr, PhD, FSRC, is Professor Emeritus (Karl Mannheim Professor for Cultural Studies) at Zeppelin University (2004–2018) and Assistant/Associate/Professor Emeritus of Sociology, University of Alberta, Edmonton, Alberta, Canada (1970–1997). He studied Economics, Sociology, Law, Social Policy Fiscal Theory and Policy at the Universities Cologne, Germany and Oregon, Eugene, OR, USA. After his PhD in 1970, he was Professor of Sociology at the Department of Sociology, University of Alberta, Canada (1979–1997), Eric-Voegelin-Professor at Ludwig-Maximilians-Universität München, Germany (1984–1985), Paul F. Lazarsfeld Professor, Human- und Sozialwissenschaftliche Fakultät, Universität Wien, Austria (2002–2003) and Alcatel Professor, TH Darmstadt, Germany (2001). Between 1977 and 2022 he held Visiting Professorships at the universities of Vienna, Zürich, Konstanz, Augsburg and Duisburg. His most recent books include *Information, Power, and Democracy* (Cambridge University Press, 2016); *Understanding Inequality: Social Costs and Benefits* (with Amanda Machin, Springer, 2016); *Knowledge: Is Knowledge Power?* (with Marion Adolf, Routledge. 2017); *Society & Climate* (with Amanda Machin, World Scientific, 2019); *Money: A Social Theory of Modernity* (with Dustin Voss, Routledge, 2020); and *Knowledge Capitalism* (Routledge, 2022). *Understanding Society and Knowledge* (Edward Elgar, 2023), and this book, *Science in Society* (with Hans von Storch, World Scientific, 2023). During the Winter Semester 2022/23, Stehr was Visiting Professor at the Universität Zürich. He is a fellow of the Royal Society of Canada and the European Academy of Arts and Science.

Hans von Storch is Director Emeritus of the Institute of Coastal Research of the Helmholtz-Zentrum Geesthacht (now: Helmholtz-Zentrum Hereon), Professor at the University of Hamburg and Professor at the Ocean University of China (Qingdao). From 1987 to 1995, he was Senior Scientist at the Max Planck Institute for Meteorology, where he worked with Klaus Hasselmann.

His research interests are climate diagnostics and statistical climatology, regional climate change and its transdisciplinary context. He has published 25 books, among them *Statistical Analysis in Climate Research* with Francis Zwiers and *Die Klimafalle* with ethnologist Werner Krauss. He has also published numerous articles. Storch is/was a member of the advisory boards of, among others, *Journal of Climate*, and the lead author of the Working Group I of the AR3 and of Working Group II of AR5 of the IPCC. He chaired the assessments of climate change knowledge for the Baltic Sea Catchment (BACC I and II) and for the Metropolitan Region of Hamburg.

Contents

Chapter 1

Introduction

1. Context

Several of the papers reprinted here take as their point of departure past reflections about the nature of the climate, climate change and its impact. Ellsworth Huntington (1876–1947) and Eduard Brückner (1862–1927) were prominent representatives of traditional scientific viewpoints, which were either fully developed climate-deterministic views or perspectives that acknowledged a weaker constraining force by climate and its variations on the economy, political power and societal well-being. These perspectives have changed in the past 50, or so, years. At the present time, when Western societies refer to future climatic conditions, a pending climate catastrophe is often the dominant narrative, especially when the assumption is that nothing significant is changing in the trajectory of societies except for the societal climate. There is a noticeable revival of the narrative of climatic determinism, but with a significant difference. In the past, the challenge was how to shield the population from climatic effects; today, the deterministic perspective has shifted to the dominant narrative protection of the climate from the impact of human activity. Some observers have tried to overcome this restrictive perception and point out that not only is climate changing but society and economy are being transformed, with new potentials and challenges. The challenge is not only to constrain climate change by reducing emissions but also to govern the adaptation of societies to climate change. So far, the global mean temperature has risen by more than 1 degree.

Another major perspective in the assembled papers concerns the use of scientific knowledge in the policy-making process and the governance of climate policies. Not only natural scientists, there are many others who are convinced that natural sciences hold the key to the solution to the climate problem. A more "extreme" group of observers is prepared to sacrifice democracy in order to assure that climate science knowledge is fully implemented (see Section 4 in Chapter 5).

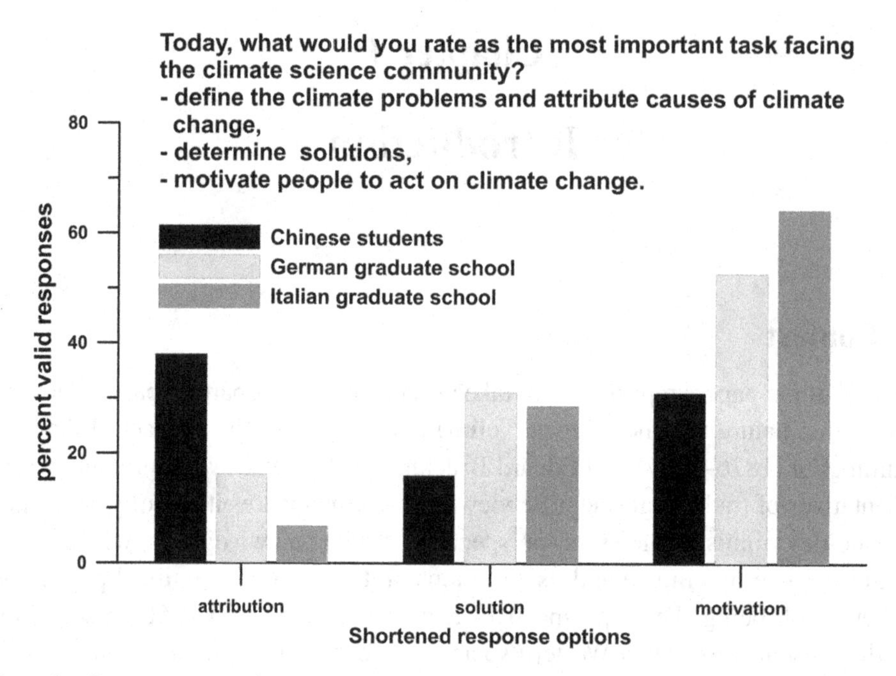

Figure 1: Responses rates of young scholars and students of different institutions, when asked what the most important question for climate science would be.

When surveying the opinions among young scholars of climate science and other fields of environmental studies, a majority of members of two graduate schools in Europe chose the answer "motivate people to act on climate change" over the alternatives "define the climate problems and attribute cause of climate change" and "determine solutions to climate change" when asked "Today, what would you rate as the most important task facing the climate science community" (see Fig. 1). Remarkably, in China, in contrast, the majority of student respondents voted for the priority of the scientific task.[1]

[1] von Storch, H., X. Chen, B. Pfau-Effinger, D. Bray, and A. Ullmann, 2019: Attitudes of young scholars in Qingdao and Hamburg about climate change and climate policy — the role of culture for the explanation of differences. *Adv. Clim. Change Res.*, 10, 158–164.

One of the practical purposes of the research carried out by Nico Stehr and Hans von Storch was to inquire into climate policies based on a comprehensive understanding of the active role/impact of humans on climate and of the climate on human societies. For our purposes, knowledge is defined as a capacity to act. Climate matters. The opposite idea views humans, human organization and entire societies as mere passive recipients to challenges issued by climatic conditions. A perspective that was valid in many regions of the world for much of historical times as climate was experienced and interpreted as a largely stable background for social, political and economic conduct. Hence, humans typically adapted by whatever means to the climates in which they happened to live. Nonetheless, climate mattered as a viable background condition. Today, and in the foreseeable future, societies have to be consciously active not only in phasing out greenhouse gas emissions but also in adapting to future climates.

The definition of knowledge as a capacity for action has several advantages.[2] For example, it implies that knowledge as a capacity for action always has multifaceted implications and consequences. The term capacity for action signals that knowledge may be left unused, may be employed for irrational ends or may be unable to be mobilized to change reality. The thesis that knowledge is invariably pushed to its limit in the absence of friction, that it is realized and implemented almost without regard for its consequences (as argued, for instance, by C.P. Snow), represents a view that is not uncommon among observers of the nature of technological development, for example. However, the notion that science and technology inherently and inevitably force their own realization in practice, by assuming such automaticity in the realization of technical and scientific knowledge, fails to give proper recognition to the context of implementation. Any conception of the immediate practical efficacy of scientific and technological knowledge (e.g., in the sense "there is nothing as practical as a good theory") overestimates the "built-in" or inherent practicality of knowledge claims fabricated in science. Suffice to say at this point that the implementation of knowledge as a capacity for action relies upon existing frameworks of social action. It is important, therefore, to base climate policy knowledge on a comprehensive understanding of society and the ways in which knowledge could become useful.

[2] The following paragraph can be found in Nico Stehr, *Moral Markets*. Boulder, Colorado: Paradigm Books, 2008: 169. More extensive discussions of knowledge as a capacity of action in Nico Stehr, *Knowledge Societies*. London: Sage, 1994, and Marian Adolf and Nico Stehr, *Knowledge*, Second Edition. London: Routledge, 2017.

2. Bridging the Two Cultures

In 1959, C.P. Snow addressed the notion of the existence of "Two Cultures" in the scientific community, one perspective guiding the humanities, and the other the natural sciences. In Snow's conception, the two cultures are at considerable odds with each other. The present book attempts to transcend Snow's dilemma and makes a case of mutually beneficial convergence of two scientific cultures of social and natural sciences ("science" in the sense of the German word *Wissenschaft* = creating knowledge).

One of the authors, Nico Stehr,[3] is a sociologist, the other, Hans von Storch,[4] a natural scientist. They met in the early 1990s, when the climate "problem" began its ascent to becoming the leading global challenge. They came from very different backgrounds —Stehr from a Northern American university campus, and von Storch from the German Max Planck Institute of Klaus Hasselmann, whose achievements were recognized by the Physics Nobel Prize in 2021.

In the early 1990s, **Nico Stehr**'s focus converged on examining, in the widest sense, the societal role of knowledge; i.e., its production, the constructive work it performs, but also the conflicts which knowledge creates. As we move through the modern world, the phenomenon we call knowledge is always involved. Whether we talk of know-how, technology, innovation, politics or education, it is the concept of knowledge that ties them all together. And still, despite its ubiquity as a modern trope, we seldom encounter knowledge in itself, we don't pay it much attention: How is it produced, where does it reside, and who owns it? Is knowledge always beneficial, will we — at some point in the future — know all there is to know? And does knowledge really equal power? Stehr was specifically interested in the topic of "practical knowledge", i.e., the sociological and philosophy of science question of the attributes that should characterize scientific knowledge claims in order to make a useful/practical difference.

Originally, Nico Stehr had the idea to write a monograph titled *Freud and Keynes*. Both John Maynard Keynes's economic theory and Sigmund Freud's psychoanalysis were considered to be outstanding scientific designs, which, moreover, had been tested in practice many times in the years since their creation: Are the instructions for action in Freud's and Keynes's theories really successful? It turned out that the practical success of Freud's theory was essentially contested, while the field of economics and economic policy widely agreed and praised

---

[3] For more information, see https://zeppelin-university.academia.edu/NicoStehr or https://de.wikipedia.org/wiki/Nico_Stehr.

[4] For further details, see http://www.hvonstorch.de/klima or https://hzg.academia.edu/HansvonStorch.

Keynes's theory as practically very promising by virtue of its effectiveness of the economic policy measures the theory implied. Keynes was stylized as the savior of capitalism. Eventually, Stehr's book was published as *Practical Knowledge*[5] in the early 1990s. Convinced that his conclusions about the leverage or the usefulness of social-scientific knowledge must also apply to natural science, Stehr set out to find a scientific discipline in which this research question could be empirically investigated. His choice fell on climate science.

In the mid-1990s, **Hans von Storch** and his group at the Max Planck Institute of Meteorology (MPI) had achieved the statistical proof of the reality of a climate change beyond the range of natural variations ("detection") and the identification of the impossibility of an explanation of this change without the explanatory factor of elevated greenhouse gas concentrations in the atmosphere (attribution).[6] His next step was to think about what climate change would mean for society and economy, with issues like downscaling[7], coastal impact and also highly simplified coupled climate–society models[8] as suggested by and in cooperation with his director Klaus Hasselmann.

The innovative and qualitatively different cooperation between Stehr and von Storch began when they met in 1992 at the newly founded Potsdam Institute for Climate Impact. The MPI scientist von Storch invited the sociologist Stehr to pay an extended visit to MPI. Stehr accepted the invitation and came in 1993 for 2 months. Both scholars began to learn the nature of the other.

One event, which displayed the differences of the two cultures, was Nico Stehr's seminar[9] at the MPI. Natural scientists are used to seeing a standing speaker, who would extensively use diagrams shown on an overhead projector — which would in most cases be more important than what is said. But Stehr did not do that. He sat down, read a manuscript, and the two overhead transparencies were

[5] Stehr, N., 1992: *Practical Knowledge*. London: Sage.

[6] Cf. Hegerl, G., H. von Storch, K. Hasselmann, B.D. Santer, U. Cubasch, and P.D. Jones, 1996: Detecting anthropogenic climate change with an optimal fingerprint method. *J. Clim.*, 9, 2281–2306.

[7] Cf. von Storch, H., 1995: Inconsistencies at the interface of climate impact studies and global climate research. *Meteor. Z.*, 4 NF, 72–80.

[8] Cf. Tahvonen, O., H. von Storch, and J. von Storch, 1994: Economic efficiency of CO2 reduction programs. *Clim. Res.*, 4, 127–141. Hasselmann, K., S. Hasselmann, R. Giering, V. Ocaña, and H. von Storch, 1997: Sensitivity study of optimal CO_2 emission paths using a simplified structural integrated assessment model (SIAM). *Clim. Change*, 37, 345–386.

[9] Stehr, N., 1993: History and Climate. https://www.academia.edu/44817135/History_and_Climate_Lecture_Max_Planck_Institute_Hamburg_Germny_May_1993.

an insignificant illustration. The audience was shocked. The bright natural scientists had a hard time to follow the arguments and theses.

The seminar conveyed an important concept, namely that of the social construct of climate and climate change, and discussed it using the example of the case of the drought in England in 1316–1319. The significance of this discussion was that two types of knowledge prevail when climate and climate change are considered in the public arena: Namely, scientific constructions and social constructions.[10]

For societal decisions, not scientifically constructed "truth" leads to decisions but a variety of social constructions consistent with culture and worldviews. For social scientists, this assertion is almost trivial, but for natural scientists, it was a kind of disenchanting assault on their claim of being the sole provider of robust knowledge. But it seems that applying this to climate science was not really common in those days — Stehr and von Storch tested that by publishing an article "Fooled by the God of the weather"[11] in a weekly news magazine and expected responses by other social scientists, who had thought along the same lines. But only a single one responded.

Now, in 2023, Stehr and von Storch have worked together for about 30 years; both are retired. They have dealt with a significant range of topics about the climate problem, the history of ideas about climate change, and the societal significance of climate and climate change. And, in fact, cooperation between social and natural scientists has become much more common in recent years as documented, for instance, by the incorporation of social sciences into the center of excellence of climate science in Hamburg, Germany.

3. Overview

This anthology reprints a series of papers, which the two authors wrote in the past 30 years. The range of essays assembled here, for the most part, have a historical emphasis and have now themselves become historical although, in many ways, our ideas continue to be topical. Our research focused on retrieving and establishing but importantly also on rejecting ideas about the nature of the interaction between society and climate and vice versa.

---

[10] In a strict sense, the scientific constructions are also social constructions, but for the sake of clarity, we look at them as different types of knowledge claims.

[11] von Storch, H. and N. Stehr, 1993: Genarrt vom Wettergott. *ZEIT*, 37, 10.9.93, 41–42.

In four chapters, different issues are touched. Since these are independently published articles, there is some unavoidable overlap between them, sometimes significantly so.

Chapter 2, on the **social construct of climate and climate change**, introduces the concept of a social construction of climate and climate change (Section 1), clarifies a role social science may play in climate science (Section 2), and finally discusses what the public understands when climate is referred to (Section 3). The last article considers the competition of socially and scientifically constructed knowledge (Section 4).

Chapter 3, on the **history of ideas of climate and climate change**, deals first with the doctrine of climatic determinism (Sections 1 and 2), then with the work of the early climate scientist Eduard Brückner who dealt at the turn of the 19th/20th century with the issues of climate change and its impact (Section 3), and demonstrates that the concept of climate change is age-old (Section 4).

Chapter 4, on the **cultures of science**, considers, first, the roles and relationship of natural and social sciences with respect to the climate issue (Section 1). Second, the different roles of scales are discussed, with the climate system being steered top-down, from large to small scales, whereas society is more of the bottom-up type, with micro-dynamics summing up to macro-dynamics (Section 2). An analysis of how physical science sees itself in advising society is given in Section 3, which has three authors, most notably the theoretical physicist Armin Bunde.

Climate policies are the topic of Chapter 5. The first two contributions deal with what can be expected to be achieved with climate policy (Section 1), and with the role of communication (Section 2). The role of science in advising sustainably the process of policy-making is considered in Section 3. The sometimes-heard claim that democratic governance and climate protection would hardly go together is deconstructed (Section 4). The last two papers, presented in Sections 5 and 6, deal with the issue of adaptation as unavoidable complimentary addition to climate mitigation policy: Commonly, this issue is simply overlooked (Section 5), or simply claimed to be the same as mitigation (Section 6).

The concluding Chapter 6 is on how the work developed — the general advice for the political process is encoded in the list of the 10 theses of the Zeppelin manifest (Section 1) from 2008. The other two sections (Sections 2 and 3) summarize what the two authors of this anthology have achieved beyond their past cooperation, with broader work on the theory of modern society, knowledge and political analysis by Nico Stehr, and more empirical analysis among scientists and the set-up of climate service by Hans von Storch. Section 4 deals with the practical implications of our work for climate servicing.

Chapter 2

Social Construct of Climate and Climate Change

In this chapter, four articles are reprinted. Section 1 reprints the first regular article of Nico Stehr and Hans von Storch, introducing the concept of a *social construct* of climate to a mostly natural science audience. Already in 1994, a first version of this paper was presented at a conference of the American Meteorological Society.[1]

Obviously, social constructions of the phenomenon of climate, competing with scientific construction of knowledge, may exert significant influence on the process of science. Thus, for any policy-relevant branch of natural science, complementary social science knowledge is needed for improving both policy-making and natural science research (Section 2).

Which societal constructions of climate change actually are currently activated in the public's mind is the subject of Section 3. The competition of the two types of knowledge about climate, climate change and its impact is considered in Section 4.

[1] Stehr, N. and H. von Storch, 1994: Climate change, the social construct of climate and climate policy. American Meteorological Society: *Proc. of the Fifth Symposium on Global Change Studies*, Nashville, Tennessee, January 23–28, 118–125.

1. The Social Construct of Climate and Climate Change*

Abstract

Different time scales of climate change and their differential perception in society are discussed. A historical examination of natural climate changes during the past millennium suggests that short-term changes, especially crucial changes, trigger a significant response in and by society. Short-term changes correspond to the 'time horizon of everyday life', that is, to a time scale from days and weeks to a few years. The currently anticipated anthropogenic climate changes, however, are expected to occur on a longer time scale. They require a response by society not on the basis of primary experience but on the basis of scientifically constructed scenarios and the ways in which such information is represented in the modern media, for example. Socio-economic impact research relies on concepts that are based on the premise of perfectly informed actors for the development of optimal adaptation strategies. In contrast to such a conception, we develop the concept of a 'social construct of climate' as decisive for the public perception of scientific knowledge about climate and for public policy on climate change. The concept is illustrated using a number of examples.

Keywords: Climate and society, Climate change, Perception of climate, Society and climate, Social construct of climate, Climate policy

1.1 Introduction

In this paper, we discuss the concept of the 'social construct of climate and climate change', its relationship to the physical climate and its impact on the design of climate policy. We illustrate our idea by comparing the present situation with historical analogs from the Middle Ages and from the first half of the present century.

In modern societies, the impact of climate and, in particular, possible future anthropogenic climate change abruptly entered public consciousness a few years ago and continues to draw considerable public attention. In the natural sciences, the view prevails that the absence of an effective response to the threat of a changing global environment results from the failure to understand the physics of ongoing natural processes. We suggest that this approach represents a flawed understanding of the dynamics of public discourse, to which problems are granted entry only as social constructs that compete for public attention with other environmental as well as social, political and economic problems. The attentiveness of

*This section originally appeared in Stehr, N. and H. von Storch: "The social construct of climate and climate change," *Clim. Res.* 5:99–105, 1995.

the public and policy-makers to such issues depends on their perceived threat to society. The required evidence for such an 'immediate threat' of climate is primarily supplied by certain extreme natural events which are independent of real climate change (such as the intense drought in 1988 in the United States or the storm season in 1993 in northern Europe).

The essay is organized as follows. First, physical aspects of climate and its natural variability and the state of our knowledge about expected anthropogenic climate change are briefly discussed. In the next section, we define the social construct of climate and (anthropogenic) climate change, and in the section following that, we deal with the dynamics of the social construct of climate by considering the interrelation between the perception of climate change, modern cultural industries and the public sphere. We then discuss the technocratic approach of designing a climate policy and contrast these ideas with past and present actual developments. Finally, in the last section, we present the options available for a more realistic and workable approach to climate policy.

1.2 The climate system and its natural variability

The physical state of the climate, and in particular the state of the lower troposphere which affects society most significantly, varies on a wide range of time scales due to various natural processes. This variability is significant for two reasons: First, it may mask any possible human-made signal (Hasselmann 1993); second, it has forced earlier societies to confront the threat of climate change, and as a result we are able to compare the response of the present-day society with the response of earlier ones.

The shortest time scales are days, with weather events such as storms or blocking events. The frequency and intensity of these events are mostly randomly distributed. There is always a chance for a '1000-year' storm to happen (cf. Hoyt 1981). The probability of such an event is small but not zero. Also, on somewhat longer time scales of weeks, droughts and floods may happen with a small but not zero probability. More precisely, the probability that at any *a priori* specified location a strong storm, a drought or a flood will happen is small. But the probability that at some location in the world there will be a strong storm, a drought or a flood is no longer small.

On time scales of years, decades and even longer periods the climate system also exhibits marked variations. The dynamics of these 'low-frequency' variations are so far not well understood, but a robust concept within which these variations appear sensible is the 'stochastic climate model' approach, which proposes the redness of the climate spectra to be a response of a slow system to short-term random forcing (Hasselmann 1976).

1.3 Anthropogenic climate change

Today, when climate change has become a household term, it is well worth reminding ourselves of the real material basis of the scenario of CO_2-induced climate change. The state of the discussion was summarized in 1990 and 1992 by the highly valued *Intergovernmental Panel on Climate Change* (IPCC), a committee made up of reputable natural scientists (Houghton *et al.* 1992). This panel concluded that there has in fact been a dramatic increase in the atmospheric concentration of radiatively active gases since industrialization and that this increase is likely to continue unless political measures are instituted to reduce emissions. Theoretical reflection as well as extensive (and expensive) experiments with detailed climate models have led to the prognosis that the increased concentration of radiatively active gases will cause a change of the global climate. Most scientists expect an increase in the overall near-surface temperature (in the range of a few tenths of a degree per decade) and an overall rise in the sea level (in the range of a few decimeters per century).

This expectation has not (yet) been unambiguously supported by observational studies because of a lack of adequate as well as sufficiently long-term and homogenous observational data. Hegerl *et al.* (1994) have shown, by using a sophisticated statistical methodology involving the use of several facets of climate model results, that the latest temperature increases are outside the expected range of natural variability and must therefore have been instigated by external factors, for instance by the greenhouse effect. This finding depends quite heavily on some estimates derived from climate models, so that the analysis by Hegerl *et al.* (1994) is rightly challenged. However, it is generally expected that the evidence to be gathered in the next years will be sufficient to attribute the observed changes to the human emission of gases and particles into the global environment.

In short, the signal produced by greenhouse warming is on the verge of emerging from the ocean of 'natural climate variability' as described in the previous section. The general near-surface warming on Earth in the last 100 years or so, with particularly high warming rates in the past few decades, is likely due to the anthropogenic greenhouse effect, both in terms of pattern and intensity. But it is possible that this 'signal' is entirely created by natural processes — and the strength of the recent signal is indeed comparable to that in the 1920s–1930s, when nobody claimed any anthropogenic climate change had occurred (see Hegerl *et al.* 1994). Because of these uncertainties, the IPCC offered the following cautionary note in its 1990 report:

> *[T]his warming is broadly consistent with predictions of climate models, but it is also of the same magnitude as natural climate variability. Thus the observed increase could be largely due to this internal variability.... [T]he unequivocal detection of the enhanced greenhouse effect from observations is not likely for a decade or more.*

It should be stressed that all global warming scenarios with some spatial details are based on 'climate models' that are the best available research tools for the study of climate variability and the design of scenarios of human-made climate change. Such climate models approximate the real climate system and are based on detailed 'general circulation models' of the ocean, the atmosphere and other components of the climate system. The oceanic and atmospheric components are relatively reliable elements in these complex climate models. Other components, such as the earth's surface or sea ice, are much less reliably represented. (For more details refer to, e.g., Washington & Parkinson 1991.)

All climate models are somehow conceptually related, not only through their basic first principles but also in their choices of how to parameterize various processes which cannot be represented directly (such as the turbulent exchange in boundary layers). Therefore, similar scenarios derived from two different climate models, say from the Geophysical Fluid Dynamics Laboratory, Princeton, NJ, USA, and the Max-Planck-Institut für Meteorologie, Hamburg, Germany, do not supply the scientific community with two independent sources of evidence that these scenarios might be correct.

Because of limited observational data, it is not possible to rigorously test the climate models in order to demonstrate that they are capable of simulating (natural and human-made) climate variability realistically. Certainly, these models have been examined with respect to weather forecasting, El Nino forecasting, the simulation of present-day climatology and other applications. Their success in doing so, together with the fact that a significant portion of the models are based on first principles, provides us with some confidence. We believe that the models describe significant sensitivities in the climate system — but we do not really know it.

1.4 The social construct of climate and climate change

Society obviously depends on climate. But what is the effect of climate anomalies on society? We claim that this dependency is largely conditional to the time scale. 'Slow' time scales of something between 1 and 30 years, which are beyond the time horizon of everyday life, are relevant for climate change, be it human-made or due to natural processes. 'Fast' time scales, which are within the time horizon of everyday life, feature 'normal extremes' such as a '100-year storm surge' and multi-year anomalies like the cold spell in Europe during the last third of the 17th century (Lindgren & Neumann 1981).

The slow variations appear to have had little social and economic impact in the past. Fast variations have produced irreversible social, economic and cultural changes either by virtue of their impact on the natural environment of a society (e.g., land lost to the sea, desertification) or by demographic (rural exodus, mortality), cultural (emerging values) and economic changes (standard of living, trade patterns, the organization and location of production, agricultural yields).

Within the context of human-made climate change the slow time scales matter. As a result, we encounter two competing images in the arena of public discourse: The (slow) climate and its changes, and the (fast) weather and climate variability (including naturally occurring rare extremes and multi-year anomalies). These two cognitive entities are physically not at all, or at most weakly, related to each other. (Climatology is just now beginning to investigate the character of the nature of interrelation of the slow and fast time scale variabilities.) Our assertion is that society is biased in its attention towards the extremes and therefore mistakes extremes as climate change.

The at times almost uncontested interpretation of climate variations by societal authorities also is an important factor in the social response to an observed real or imaginary climate change. Such authorities may be scientists or charlatans but also the modern media, superstition, or religious institutions. Another important factor is, at any given time, the competition for public attention and solutions among contemporaneous social problems. There are many more or less urgent social problems which compete with the threat of climate change for scarce public attention and resources (e.g., Ungar 1992). Because of these processes, the public never obtains a perspective of climate as elaborated by the physical experts in an unmediated fashion but only a filtered image of it, namely the social construct of climate. We suggest that the climate and its social construct can be independent entities or events.

1.5 Cultural industries, the public sphere and the perception of climate change

Climatologists concerned with their scientific capacity to reliably derive scenarios of climate change have devoted a significant part of their literature to the overall problem of the uncertainties that surround such scenarios now and in general.

In spite of this generally cautious stand, there are climatologists who stress the risks for society of responding to such contestable conjectures with incredulity (e.g., Schneider & Mesirow 1976, Kellogg 1978).

However, neither the manufacture nor the communication of research on climate and climate change occurs in a social vacuum. These activities both within and external to science are linked to various social practices ordered across time and space. In the following, we will only touch on two major aspects of these social practices, which affect the ways in which scientific conjectures can be communicated to the public without encountering disinterest or disbelief. Particularly the communication of research findings to individuals and groups outside of the scientific community is affected significantly by social processes that influence the organization of images and ways in which people make sense of the dynamics of society and natural processes. Research results, however, carefully constructed, are filtered and transformed in various patterned ways. These processes, for example, those within cultural industries as one of the major sources of information and meaning in modern society, mediate and reconstruct scientific findings generated by climate researchers. And the output of these processes will determine how these findings ultimately are interpreted by the public at large as well as by groups such as social movements (cf. Lowe & Morrison 1984, Lacey & Longman 1993, Singer & Endreny 1993).

Cultural industries do not merely provide access to the broadest range of information, reasoned advice and interpretative analysis on various topics in order to facilitate rational discussion and public decision (as might be the case in some ideal sense). At best, such a conception of the communication sector of modern society provides for a desirable yardstick against which its actual performance may be assessed. Since cultural industries are subject to various other significant constraints, not least among them economic and ideological pressures, their actual performance, the range and depth of coverage, most often does not sustain and support a public sphere in which research findings are represented and reinterpreted in a manner that prevent them from being radically transformed and in which cautionary notes and qualifications are completely ignored (see Gamson 1993).

In short, the desire of climatologists to communicate their findings in an unequivocal fashion for public discourse encounters, first, the obstacle of modern cultural industries and their peculiar contingencies. Second, the public or, better, various segments of the public interpret the research findings in ways which may or may not correspond to the intentions of the researchers.

When researchers have examined the public response to social and environmental problems (such as drug use or public encounters with diseases or disasters), two perspectives have dominated the study of the collective or individual response to social issues, namely the objectivist and, the constructivist approaches (cf. Merton & Nisbet 1966, Douglas & Wildavsky 1982, Douglas 1992).

The objectivist approach is based on the premise that the 'threat' in question, for example, climate change, is quite real, can be demonstrated scientifically and likely will cause serious harm to human life and society. The constructivist approach, in contrast, concentrates on the public perception of the risks and emphasizes the ways in which perception and assessment of risks are influenced by social and cultural factors. The first view tends to focus on the manufacture (by experts) of conjectures, stressing their objective, undeniable consequences, while the second approach chooses to emphasize the reception (by laymen) of such hypotheses in different social, cultural and political contexts. The disparities and discrepancies between the two forms of inference often are highlighted and lead to the conclusion that a generally 'true' definition of risks and threats is at best a dubious undertaking (e.g., Rayner & Cantor 1987).

In general, however, research into risk communication, the public perception of social problems and threats posed by various natural or social events takes place at a pragmatic middle ground between the two extremes, denying neither that threats are objective nor that the public response may at times vary considerably and/or chose to ignore such warnings altogether (e.g., Goode 1989). Lacey & Longman (1993, p. 239), for example, conclude in their analysis of recent press coverage in England of environmental and development issues, 'The coverage of global warming which was at a height in 1989 and 1990 had dwindled to almost pre-autumn 1988 levels by spring 1991. This was a general phenomenon across the range of newspapers considered. This is a disturbing feature of press coverage. It means that despite the worsening of the actual condition (greenhouse gas emissions and global warming) and no indication that public concerns had diminished the press "gatekeepers" have decided that the issue is no longer newsworthy.' The literature on risk communication often aims to find ways of more effectively communicating 'objective' expert information to a public whose perception is 'mediated' by various cultural processes (cf. Wiedemann 1991).

However, such an approach to the communication and public perception of risks simplifies — as we already have tried to demonstrate in a general sense — complicated matters to a considerable extent. To begin with, the communication of information, especially the extent to which conjectures are 'believed', is rarely a matter of the 'quality' (for instance, in the sense of the objectivity or scientificity) of its contents. On the contrary, the quality of rapidly changing or enduring social relations and of salient cultural resources (e.g., world views) among active agents in science, cultural industries and the public sphere matter more. The formulation of risks or the assessment of hazards, the reporting and interpretation of these issues and, last but not least, the public response to these accounts involves at each juncture active agents with distinct purposes engaged in arriving at their interpretations of what then become socially constructed meanings. The outcome is a complicated form of discourse and intersection of social contexts that cannot easily be influenced to assure that a specific framing of the issues and of the conclusions prevails. The manufacture of and subsequent response to objective claims formulated in statistical terms, for example, and generated by climate research is no exception. Climatologists construct objective conjectures about climate according to certain social practices and standards prevailing and accepted in the scientific community generally. Objective and constructivist features of scientific scenarios and expected risks engendered by climate change tend to mix and blend into each other (cf. Krohn & Krücken 1993), and this is especially the case as the scenarios enter the arena of public discussion in the form of 'predictions'.

1.6 Climate policy

The significant implication of distinguishing between climate and its social construction is that it is only the social construct which ultimately shapes climate policy, whereas the climate itself plays no or only an insignificant role in the process of designing a climate policy.

In the scientific community, economic concepts and perspectives have dominated in the discussions on how to respond to the possibility of human-made climate change. And in the intellectual tradition of neo-classical economics, a perfectly informed society is expected to design an 'optimal' response strategy (e.g., Nordhaus 1991, Tahvonen *et al.* 1994). A prototype would be a schematic depiction of the relationship between the global environment and society in which two entities, 'climate' and the 'socio-economy', are assumed to interact with each other via environmental parameters, such as temperature or precipitation (which, in turn, affect the biosphere and thus man), and the emission of radiatively active gases. The 'costs' of a climate change ('damage' or 'adaptation' costs) as well as the costs of changing the economy required to avoid or diminish climate change ('abatement' costs) are, at least in principle, known and can be quantified (in money or moral units). This quantification is done according to social norms and political decisions that represent societal preference and utility scales. An 'optimal' climate policy is then designed to minimize the total costs taking into account the damage costs and the abatement costs (Hasselmann 1990).

We would like to contrast such a viewpoint — that best is called the 'technocratic' approach — with a perspective according to which it is not climate itself but the social construct of climate that is the dominant factor. We suggest that society has failed to pay sufficient attention to the real and thus slow climate-change signal. Instead, society responds to the social construct of climate change, thereby mistaking natural extremes as indicators of climate change. We illustrate our conception by way of three examples.

Between 1315 and 1317, the harvest failed in England, mainly because of persistent summer precipitation. As a consequence, there was a famine and diseases spread (killing up to 10% of the population). The authorities, essentially the Church, had warned its subjects prior to the failed harvests that God would punish them if they did not adopt higher moral standards in their life. The climatic extreme that occurred was interpreted as a climate change. The (only) believable factor controlling climate was said to be God, and the change in the climate thus reflected God's anger and revenge. Because of the life-threatening character of the implications of climate change (famine, death), 'adaptation' was not an acceptable climate policy. The only available option was 'abatement', which meant to put an end to God's wrath. And exactly that was the social response at the time. As one social historian (Kershaw 1973) reports:

> *[T]he Archbishop of Canterbury ordered the clergy to perform solemn, barefooted processions bearing the Sacrament and relics, accompanied by the ringing of the bells, chanting of the litany, and the celebration of the mass. This was in the hope of encouraging the people to atone for their sins and appease the wrath of God by prayer, fasting, almsgiving, and other charitable work.*

This climate policy was perceived as successful: The climate anomaly disappeared; the harvests recovered. The social construct of climate and climate were clearly unrelated in this case. Another example of a social construction of climate change during the Middle Ages might make reference to witches who were widely perceived to modify climate either directly through witchcraft or indirectly by causing God's anger about failure to take action against the evil practice of the witches (Behringer 1988).

The idea that emissions of greenhouse gases might artificially change the global climate, with an increase in the near-surface temperature, was already proposed in the late 19th century by the Swedish scientist Arrhenius (1896). For many years this notion was considered an intellectually appealing but practically unimportant thought. Only in the 1970s were the possible impacts of the anthropogenic greenhouse effect discussed more seriously. In the 1980s the 'greenhouse effect' became the most important topic in climate research, with increasing funding ever since. The public suddenly appeared to accept the greenhouse problem as a significant issue in the aftermath of several extreme events.

The North American drought in 1988 was crucial in North America. During a hearing of the United States Senate, the well-known climate researcher James Hansen declared the drought with '99 percent certainty' to be related to the anthropogenic climate change (Schneider 1989). This statement had a poor substantive basis and appears dubious in view of the absence of further droughts in subsequent years (the headlines in the summer of 1993 were dominated by reports of dramatic flooding in the same regions). An alternative explanation for the drought, that it resulted from a response to a peculiar configuration of sea-surface temperature anomalies in the North Pacific, was put forward by other climatologists (Trenberth *et al.* 1988).

In the spring of both 1991 and 1993, northern Europe experienced a series of severe storms which caused significant damage. The storms were interpreted by the media as an indicator of the predicted climate change, and even reputable scientists declared more or less openly that the frequency of intense storms had increased and would further increase as a response to human emissions of greenhouse gases. A statistical analysis of the frequencies of storms in the North Sea area and other parts of the North Atlantic area (von Storch *et al.* 1993) in the past 100 years revealed no such systematic changes. The result was largely disregarded by the media even though a short version was published in Nature (Schmidt & von Storch 1993).

A further example refers to the decades of the 1920s and 1930s in the Northern Hemisphere. Within two decades, from 1911–1920 to 1931–1940, the annual mean temperature in the Northern Hemisphere increased by 0.3°C. Local changes were as high as 1°C and more. The public did not take notice of this change, although its magnitude was comparable to the present change (the Northern Hemisphere's mean temperature change from the 1971–1980 decade to the 1981–1990 decade was only 0.25°C according to the most reliable estimates). We suggest that climate change in those years simply failed to become a major public concern because of the competition posed by traumatic social problems such as the societal reorganization after the First World War, the economic depression and the formation of totalitarian regimes.

1.7 Options for climate policy

We therefore argue that any climate policy is subject to the following dilemmas:

— If slow climate change takes place and the public has been warned of such a change by the political and/or scientific authorities, then the real slowly evolving signal will hardly be noticed. Instead, the public will accept extremes, which are consistent with (but in fact mostly unrelated to) the warnings, as 'proofs' of the reality of climate change. An active abatement or adaptation policy can be designed, but whether this policy will be adequate remains an open question.

— If the climate changes gradually and the public is not concerned about such a change, passive adaptation will take place. The naturally occurring extremes are then accepted by the public as unavoidable natural interruptions.
— If climate does not change, but the public nevertheless expects a climate change, then any extreme (or multi-year anomaly) will be interpreted as evidence of the climate change and a climate policy will be instituted according to the norms accepted in a given society and historical period.
— If the climate is stationary and the society does not expect changes, extremes will create no demand for a climate policy.

The last configuration is the most frequent in history. In most historical writings, the weather and weather-related catastrophes are discussed mostly for reasons of completeness (e.g., Weikinn 1958–61). The case 'England 1314–1317' belongs to the third category, the case 'Northern Hemisphere 1920–1930' must be assigned to the second category and the present situation may belong to the first or the third groups.

We conclude that:

— The physics of climate change is largely incomprehensible to the public. The anticipated climate change occurs on time scales much longer than the 'time horizon of everyday life' so that people must respond to threats they actually do not experience personally. Even social groups closely dependent on environmental factors sensitive to climate change, such as the agricultural sector or people living in coastal areas, find it difficult to deal with a slow but steady climate change.
— The notions of climate and social construct of climate and climate change, while not contradictory, are often independent of each other.
— A 'reasonable' societal reaction to the climate change induced by humans, which, at least in principle, can be controlled by political measures, cannot realistically be expected. Such a reaction perhaps could be produced by a skillful manipulation of the 'misunderstanding' of extremes (it appears that such an option does exist in the minds of some natural scientists) or by way of a vigorous public campaign.

Research on global environmental change in general and on climate change in particular is still widely considered essentially a subject area of the natural sciences. Despite the recent well-publicized critique of the notion of global warming by a minority of climate researchers and others (e.g., Salmon 1993), the establishment and representation of consensus on climate change by the natural sciences (IPCC 1990, Houghton *et al.* 1992) continues to exert the most influence in the international political arena. But it also is necessary to critically examine the notion that the scientific authority of knowledge on climate change is somehow natural rather than constructed within the scientific community.

In any case, there is a substantial need for interaction between the strictly separated and persistently sovereign domains of natural science climate research and social research in order to understand the interdependencies between climate and the social construction of climate. We need, for example, more historical comparisons with present situations. In addition, empirical analyses of the societal perception of climate and weather are required to begin to answer key questions, such as 'What is so special about the climate problem that it at times appears to be more serious than most other social problems?'

Literature cited

Arrhenius SA (1896) On the influence of carbonic acid in the air upon the temperature of the ground. Phil Mag J Sci 41:237–276.

Behringer W (1988) Hexenverfolgung in Bayern. R Oldenbourg Verlag, München.

Douglas M (1992) Risk and blame. Essays in cultural theory. Routledge, London.

Douglas M, Wildavsky A (1982) Risk and culture. An essay on the selection of technological and environmental dangers. University of California Press, Berkeley.

Gamson WA (1993) Talking politics. Cambridge University Press, Cambridge.

Goode E (1989) The American drug panic of the 1980s: Social construction or objective threat? Violenc Aggress Terror 3:327–348.

Hasselmann K (1976) Stochastic climate models. Part I. Theory. Tellus 28:473–484.

Hasselmann K (1990) How well can we predict the climate crisis? In: Siebert H (ed.) Environmental scarcity — the international dimension. JCB Mohr, Tübingen, pp. 165–183.

Hasselmann K (1993) Optimal fingerprints for the detection of time-dependent climate change. J Clim 6:1957–1971.

Hegerl GC, von Storch H, Hasselmann K, Santer BD, Cubasch U, Jones PD (1994) Detecting anthropogenic climate change with an optimal fingerprint method. Max-Planck-Institut für Meteorologie Rep 142, Hamburg.

Houghton JT, Callander BA, Varney SK (1992) Climate Change 1992. The supplementary report to the IPCC Scientific Assessment. Cambridge University Press, Cambridge.

Hoyt DV (1981) Weather 'records' and climate change. Clim Change 3:243–249.
IPCC (1990) Scientific assessment of climate change. The policymakers' summary of the Report of Working Group I to the Intergovernmental Panel on Climate Change. World Meteorological Organization/United Nations Environment Programme, Geneva.
Kellogg WW (1978) Global influences of mankind on the climate. In: Gribbin J (ed.) Climatic change. Cambridge University Press, Cambridge, pp. 205–227.
Kershaw I (1973) The great famine and agrarian crisis in England, 1315–1322. Past Present 59:3–50.
Krohn W, Krücken G (1993) Risiko als Konstruktion und Wirklichkeit. In: Krohn W, Krücken G (eds) Riskante Technologien: Reflexion und Regulation. Einführung in die sozial-wissenschaftliche Risikoforschung. Suhrkamp, Frankfurt am Main, pp. 9–44.
Lacey C, Longman D (1993) The press and public access to the environment and development debate. Sociol Rev 41:207–243.
Lindgren S, Neumann J (1981) The cold and wet year 1695 — a contemporary German account. Clim Change 3:173–187.
Lowe P, Morrison D (1984) Bad news or good news: Environmental politics and the mass media. Sociol Rev 32:75–90.
Merton RK, Nisbet R (eds) (1966) Contemporary social problems. Harcourt Brace & World, New York.
Nordhaus WD (1991) To slow or not to slow: The economics of the greenhouse effect. Econ J 101:920–937.
Rayner S, Cantor R (1987) How fair is safe enough? The cultural approach to societal technology. Risk Anal 7:3–9.
Salmon J (1993) Greenhouse anxiety. Commentary 45:25–28.
Schmidt H, von Storch H (1993) German Bight storms analysed. Nature 365:791.
Schneider S (1989) Global warming: Are we entering the greenhouse century? Sierra Club Books, San Francisco.
Schneider SH, Mesirow LE (1976) The Genesis Strategy: Climate and survival. Plenum Press, New York.
Singer E, Endreny PM (1993) Reporting on risk. How the mass media portray accidents, diseases, disasters, and other hazards. Russell Sage Foundation, New York.
Tahvonen O, von Storch H, von Storch J (1994) Economic efficiency of CO_2 reduction programs. Clim Res 4:127–141.
Trenberth K, Branstator GW, Arkin PA (1988) Origins of the 1988 North American drought. Science 242:1540–1645.
Ungar S (1992) The rise and relative fall of global warming as a social problem. Sociol Quart 33:483–501.
von Storch H, Guddal J, Iden K, Jonsson T, Perlwitz J, Reistad M, de Ronde J, Schmidt H, Zorita E (1993) Changing statistics of storms in the North Atlantic? Max-Planck-Institut für Meteorologie Rep 116, Hamburg.

Washington WM and Parkinson CL (1986) An introduction to three-dimensional climate modelling. University Science Books, Mill Valley, CA.

Weikinn C (1958–1961) Quellentexte zur Witterungsgeschichte Europas von der Zeitwende bis zum Jahre 1850, 4 Volumes. Akademie Verlag, Berlin.

Wiedemann PM (1991) Strategien der Risiko-Kommunikation und ihre Probleme. In: Jungermann H, Rohrmann B, Wiedemann PM (eds.) Risikokontroversen. Konzepte, Konflikte, Kommunikation. Springer-Verlag, Berlin.

2. The Case for the Social Sciences in Climate Research†

Abstract

The present dilemmas brought about by anthropogenic climate change are in many ways unprecedented. Knowledge about the physical nature of global climate changes is not sufficient to move from comprehension to a solution of the problem. The historical record shows that past generations, too, have been fascinated and concerned about the impact of climate as well as anthropogenic climate change on society. But these efforts have, for the most part, been informed by the doctrine of climate determinism.

We ask therefore what a more realistic form of impact research, as a basis for climate policy, must look like. We argue that the conception of the issue as an "optimal control problem" is inadequate. Impact research has to be cognizant of the dynamic social construct of climate. As a result, climate policies as a form of managed climate change have to draw extensively on social science expertise.

2.1 Introduction

Climate research has thrived within the scientific community for the past decades. To date, climate research has dealt mainly with questions about the physical dynamics of climate understood as a natural phenomenon. For the purposes of policy, accurate numerical and system-analytical answers are considered sufficient answers while the translation of such knowledge into practical decisions in the societal and political realm is taken for granted.

But the success of climate research has not led to the institution of policies by balancing expected damages and abatement costs to mitigate, or even avoid, the detrimental consequences of expected anthropogenic climate change. Instead, the — often misinterpreted — information provided by climate research is responsible for the creation of alarm ("climate catastrophe") among the public and political inactivity. In everyday life, the magic terms "greenhouse effect and global warming" are now widely known; but equally widespread is confusion about the nature of these concepts. Political actions are mostly limited to verbal announcements and more or less generous funding of climate research.

Natural scientists continue to be as optimistic and well-meaning as most natural scientists have been in the past. To avoid misconception, we consider it

†This section originally appeared in von Storch, H. and N. Stehr: "The case for the social sciences in climate research," *Ambio* 26:66–71, 1997.

imperative that social science expertise is brought into the center of climate research. We present a series of cases which demonstrate how social science expertise can help build a more holistic and realistic view of climate and society.

In this section, we take a skeptical stance towards the relevance of "natural" scientific information about climate and climate change for society. Such skepticism does not imply that we question the reality of anthropogenic global warming (compare references 1–3). However, the existence and comprehension of the natural process does not necessarily imply its relevance for society. Whether anthropogenic climate change is socially relevant or not, has to be explored by climate impact research, a field attracting more and more interest. In the following two sections, we question two conventional approaches pursued by climate impact research.

Specifically, we address two questions:

— *Is climate impact research a new scientific field?* We will show that it is not new but a forgotten "science" that has fascinated countless generations in many societies. However, past climate impact research was mainly of the "climatic determinism" genre, a paradigm which disappeared from the scientific discourse perhaps because of its intimate relation to racial theories. But even if it disappeared from the scientific agenda, climate determinism is still a most vivid concept among the public in contemporary society, and that includes decision makers and politicians.

— *Is it sensible to consider the social consequences of global warming as an "optimal control problem" which requires the construction of "climate policy" that balances expected abatement costs against expected climate change damage costs?* We will assert that this approach is questionable because it disregards the dynamics of social value attribution over time. The costs of climate change perceived by future generations may be radically different from our present measures of value, or social preferences.

While these two questions address the conventional climate impact research pursued by geographers, ecologists and economists, we see the need for another type of climate impact research which has to do with the public perceptions and beliefs, with the subjective role of natural scientists and decision makers and their interaction with society. In line with this general point, we ask in a later section: Is the public's perception of global warming consistent with the views of natural scientists? What is the contemporary social construct of climate and climate change? We will demonstrate that this social construct as expressed by the public at large and reinforced as well as dramatized by the media, is often far removed from what natural scientists consider to be the case.

The section concludes with a preliminary list of research questions which are not only intellectually appealing, but also of relevance for dealing with the threat, or scare, of global warming.

2.2 Climate impact research

Independently, whether we accept anthropogenic global warming as a reality or, to some degree, as a possible evolution which may take place in the future, the practical implications need to be explored. Thus, climate impact becomes a key research task. We have to ask to what extent and how climate and climate change determine the performance of natural and managed ecosystems and economic and social structures and how any mitigating efforts, i.e., costs associated with the management of climate change, in turn affects society.

One avenue of inquiry of climate impact research, evident from the earliest time of civilization, has been the speculation about the effect of climate on humans. For example, in classical Greece, Hippocrates, suggested in his treatise on "Air, water and places," that knowledge about climate ought to be used to explain the psychology and physiology of humans. The differences in habits of life and character between East and West were thought to be a result of the differences in climate. During the enlightenment, the educated part of the population of France, Germany and England, spent enormous intellectual energy arguing about the climatic determinants of the civilizational peculiarities of entire nations. Philosophers such as Montesquieu in his influential "Esprit des Lois" and Herder in his "Ideen zur Philosophie der Geschichte der Menschheit" advanced widely discussed ideas about the significant constraint that climate represents.

But even in our century, climatic explanations of history and the theory of significant climatic influences on individuals and societies have flourished. While earlier speculations about the impact of climate were largely derived from casual observation, the American geographer Ellsworth Huntington introduced the quantitative method. In his monograph "Civilization and Climate" (note reference 4), Huntington advanced the hypothesis, widely accepted by the public and appreciated by fellow scientists, that the formation of a civilization would be possible only in areas where favorable climatic conditions prevail. His conclusions were based on a statistical analysis of the work records of factory workers and marks of college students. Huntington claimed to have shown that humans are most energetic and productive at a temperature of c. 15–21°C, as well a moderate annual range of temperature and the presence of short-term variability. The latter was thought to be stimulating both in terms of mental and physical energy and health. Not surprisingly, such climatic conditions prevail in modern times in Western and

Central Europe, most of North America, to some extent in Japan, and in Australia and some parts of southern South America. Conversely, Huntington claims that both physical and mental activity decline with extremes of either heat or cold. As a verification of his hypothesis, Huntington showed two maps (Fig. 1) displaying the distribution of health and energy as derived from climatic conditions, and the distribution of civilization, as determined by a survey among experts.

Not surprisingly, similar ideas were in fashion in Nazi Germany where the social psychologist Willy Hellpach for instance wrote in an essay entitled "Culture and Climate," published as part of a volume on the general topic "Klima-Wetter-Mensch":

> *Prevalent in the North ... are the character traits of sobriety, harshness, restraint, imperturbability, readiness of exertion, patience, stamina, rigidity, and the resolute employment of reason and determination. The prevalent traits of the South are liveliness, excitability, impulsiveness, engagement with the spheres of feelings and imagination, a phlegmatic going-with-the-flow or momentary flare-ups. Within a nation, the northerners are more practical, reliable, but inaccessible, and the southerners devoted to fine arts, accessible (sociable, likable, talkative), but unreliable (5, 6).*

After racial theories, which are essentially a kind of racial determinism, were discredited, their intellectual siblings' geographic determinism and climatic determinism were also rendered obsolete in the social sciences. Today, the incorporation of environmentally determined impacts on human behavior is almost considered taboo within most social science discourse. In the natural sciences, the concept has survived, to some extent. Such a scientific perspective is pursued by biometeorologists, investigating, for example, the effects of heat waves on domestic violence or mortality rates.

In the natural and social sciences, the relation between climate, social conduct, attitudes and abilities is examined if at all in a most cautious manner, The general public still accepts the concept of climatic determinism as illustrated for example by an article in the journal *Weather* (7), in which the author claims:

> *... on apparent correlations between the character of the people of a region and the climate prevailing there ... intolerant acts have often been conducted by people from areas in mid-latitudes where seasonal temperature extremes are large.... In the 1930s, fascism took over in Spain, Germany, Italy and Austria; all [have a seasonal temperature range] about 20 deg C. ... It may never be possible to prove absolutely that a mild climate in mid-latitudes helps to foster a tolerant society or that an extreme climate may predispose people towards intolerance*

> *... if this is recognized it could help to identify potential problem areas in the field of human relations so that timely action can be taken to mitigate threats to peace. ... Perhaps the absence of seasonal change helps to foster a relaxed attitude because there is no need to make elaborate plans to cope with the rigors of a cold — winter and/or a very hot summer. However, where [the seasonal temperature range] is large, the pace of life is driven by the seasons, enforcing discipline of timely preparation for the extremes; here, less relaxed mental attitudes may develop.*

Also, a survey among college students from 26 countries conducted by Pennebaker *et al.* (8) finds support for the persistent resonance among the young and educated segment of the population for Hellpach's ideas and, therefore, for the stereotypical image of different Northern and Southern personality types.

In decision-making processes on climate matters and in possible conflict with the hard information provided by natural sciences, one should not underestimate the relevance of the widespread belief in "climatic determinism" as well as other relevant cultural climate-related doctrines, for example, the notion that climate is constant.

Another course of inquiry of earlier climate, and climate impact research has concerned the variations or changes in climate. Attentive observers detected already in the 18th century that climate is not constant, and researchers speculated about the reasons for such changes. As a result, the dichotomy of natural and anthropogenic climate change was introduced. In 1770, the American physician Williamson described a change in climatic conditions in the North American colonies, and linked this favorable change to the ongoing settlement that produced increased drainage and deforestation (9). Similarly, the saying "The rain follows the plough" describes the idea of a beneficial climate change caused by the transformation of the North American prairies into agriculturally managed farmland.

In the 19th century, widespread discussions took place in Europe, Australia and North America, about climate change due to deforestation and, sometimes, reforestation (10, 11). This debate was not confined to the scientific community of the day but found considerable echo in the media and in politics. Moreover, the discussion was rather similar to the present one about the interpretation of the current warming trend, that is: Are we faced with just another long-term swing in the course of natural variability or is it becoming warmer because of anthropogenic modifications of the environment?

The conviction that the changes were anthropogenic led in several countries to the establishment of governmental and parliamentarian committees for the purpose of designing proper response strategies.

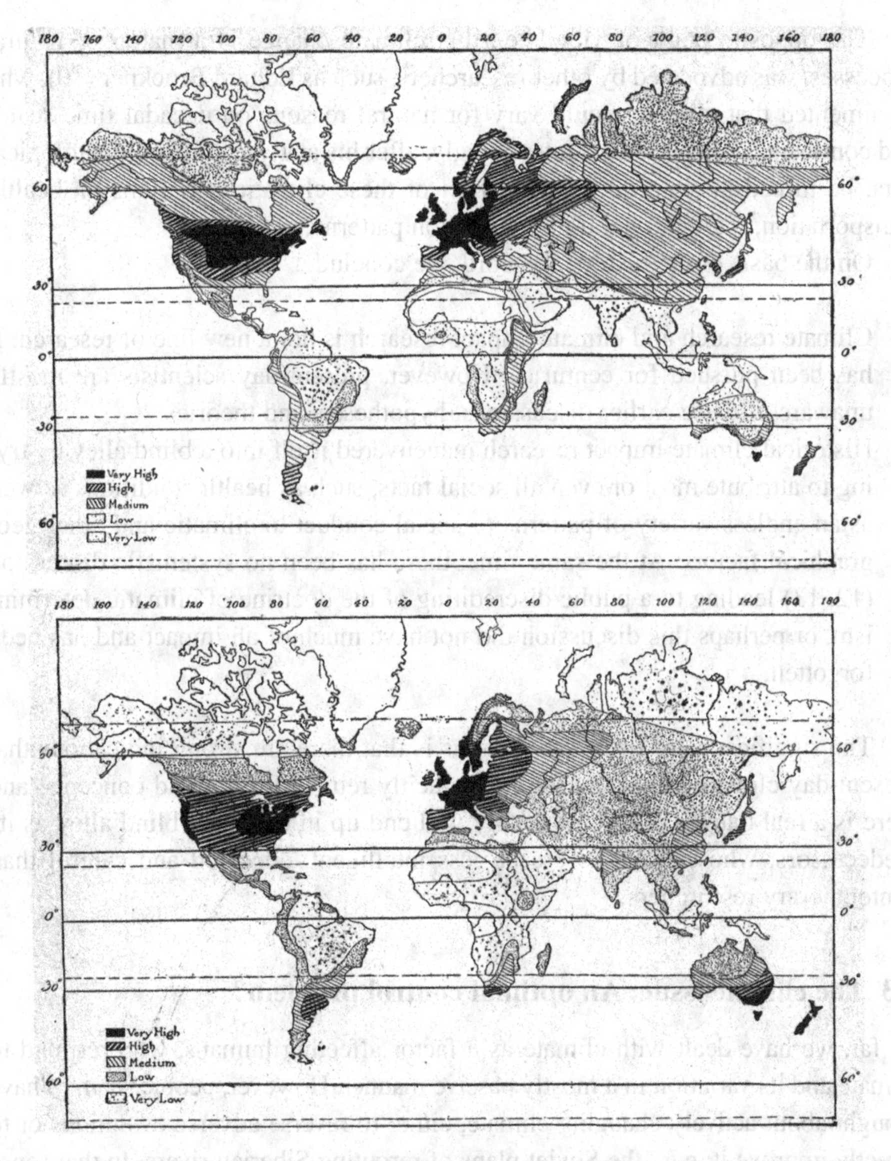

Figure 1: Huntington's (4) key argument for his "climate hypothesis of civilisation." Top. Huntington's analysis of climatically determined "health and energy," and bottom: The distribution of "civilisation" derived from a survey among contemporary "experts." From the two maps, Huntington concluded that favorable climatic conditions would be a necessary condition for a civilisation to form.

The opposite point of view, i.e., that climate change is a matter of natural processes, was advocated by other researchers, such as Eduard Bruckner (10), who documented that climate would vary for natural reasons on decadal time scales and continental spatial scales. Interestingly, after his analysis of the climatological data, he turned his interest to the impact of these climatic variations on health, transportation, international trade, migration patterns, etc.

On the basis of the historical record, we conclude:

1. Climate research and climate impact research is not a new line of research. It has been pursued for centuries. However, present-day scientists are mostly unaware of these earlier discussions, hypotheses, and theories.
2. Historical climate impact research maneuvered itself into a blind alley by trying to attribute most or even all social facts, such as health conditions as well as an endless variety of patterns of social conduct to climatic and other geographical factors. At the same time, there has been no systematic discussion (12, 13) leading to a public discrediting of the doctrine of climatic determinism, or perhaps this discussion did not have much of an impact and has been forgotten.

The significance of these conclusions is that there are strong indications that present-day climate impact research has tacitly returned to the old concepts, and there is a real danger that it eventually will end up in the same blind alley as its predecessors, who certainly were no less intelligent, educated and careful than contemporary researchers.

2.3 The climate issue: An optimal control problem?

So far, we have dealt with climate as a factor affecting humans, who respond to climate and its variation in a mostly passive manner. However, people seem to have thought about actively changing climate, either to reverse adverse evolutions or to directly improve it; e.g., the Soviet plans of rerouting Siberian rivers. In that sense at least, there is a history of managed and even planned climate change. In the case of the anthropogenic greenhouse effect, most members of society and governments consider it a worthwhile goal to limit the expected anthropogenic climate change, in order to ensure that expected damages remain within acceptable bounds.

From a macro-economic perspective, climate change may be understood as a situation in which the creation of economic welfare has the secondary effect of causing damage to the environment. In the case of anthropogenic climate change, the harmful side effects are, for example, damages such as rising sea levels. These damages create the need for a number of adaptation measures; e.g., the construction of dikes which exploit economic resources that could alternatively be used for the creation of welfare.

The problem is related to the tragedy of the commons (14): All actors together exhaust a common resource, namely the atmosphere as a dump for gaseous by-products of energy generation. By doing so, individual profits are gained. The effect for the common good, however, results in adverse effects for everybody, independently of the amount of emissions by each individual.

Assuming no intervention at all, economists expect a monotone increase of greenhouse gas emissions, the so-called "business as usual" policy. The alternative would be that the world's governments agree on a joint policy aiming at limiting damages on the basis of regulating emissions. The social optimum would be an emission plan for the entire world, balancing the costs associated with the reduction of emissions with the expected cost of damages in the foreseeable future. In strictly economic terms, a time-dependent emission path is aimed for, so that the marginal abatement costs equal the marginal adaptation costs. This idea was pioneered in economics by Nordhaus and in climate research by Hasselmann (15–18). Cast in these terms, the climate problem reduces itself to an optimal control problem, with the emission path as control variable and climatic conditions as state variables (19).

Hasselmann has condensed his approach into the Global Environment and Society (GES) model, in which two dynamical entities, namely the climate system and the economic system interact (Fig. 2). The economic system affects the climate system by wastes such as carbon dioxide (CO_2), and the climate systems respond with a change of, say, sea level. Any waste reduction is associated with costs. Climate changes incur costs as well. The role of public policy is to minimize the total costs, the exact measure of which is left to society.

This optimal control approach is not only intellectually tempting but may also appeal to policy makers. Undoubtedly, it represents a rewarding and informative perspective for discussing the problem at hand. On the other hand, it functions only on the basis of various assumptions, some of which are not explicitly stated. Some simplifications, such as the absence of natural climate variability, could easily be accounted for by modifying the involved dynamical models. Other static assumptions are more difficult to justify. For example, a major assumption of the model is that future generations will accept our values and our concept of a healthy environment. The macro-economic models assume that the assignment of value is mostly constant, perhaps with a discounting element, but without a significant change in the relative designation of values for, say, healthy forests and religious prescriptions. But we know that societal values undergo complex and barely

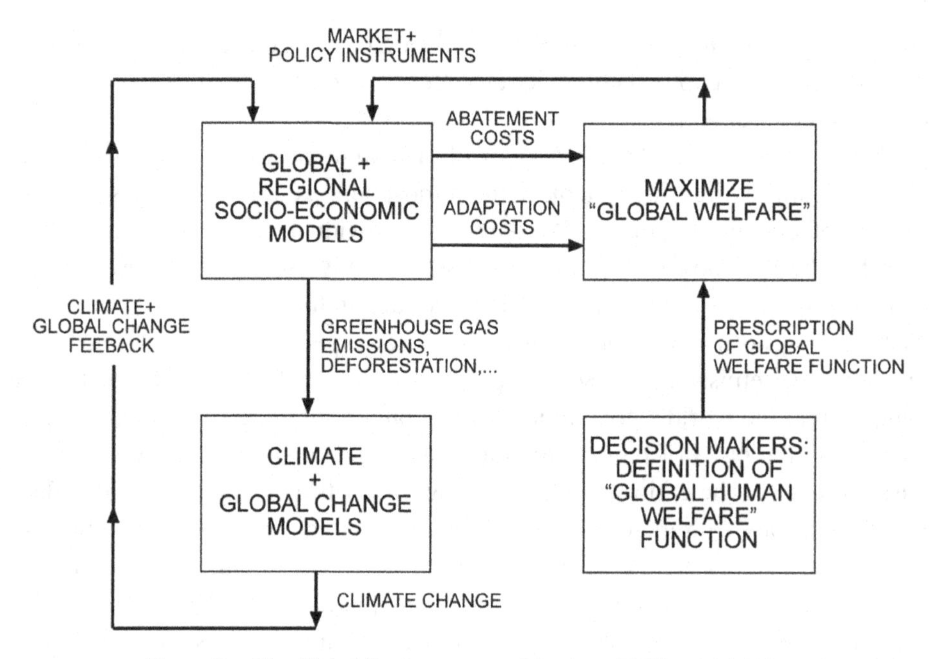

Figure 2: The Global Environment and Society (GES) model (16).

predictable transformations. What is of utmost relevance for significant segments of the general public today, may be irrelevant only a few years let alone decades hence. In other words, models like GES lack a module describing the dynamics of social value assignment. Given the state of our knowledge, it is hardly imaginable that such models may be reliably set up to be used in integrated assessment models.

To illustrate the general point, we offer an example from medieval times (compare to Section 1 of this chapter). In the years 1315 to 1319, parts of Europe suffered from severe weather-induced shortages of food; the problem was severe in England, among other countries. The reconstructed air-pressure distribution for the summer 1315 illustrates the situation (Fig. 3). A persistent anomalous cyclonic circulation over Central Europe brought unusual cold and rainy conditions for the summer with disastrous consequences for the harvests.

The hostile climatic conditions were interpreted by the contemporary authorities, i.e., the church, as a control problem. The adverse climate was seen and understood as being brought upon society by God in response to sinful conduct. In a sense, society was confronted with anthropogenic climate change. Any business-as-usual response would be associated with unbearably high damage costs — famine, epidemics, high mortality apart from unfavorable perspectives

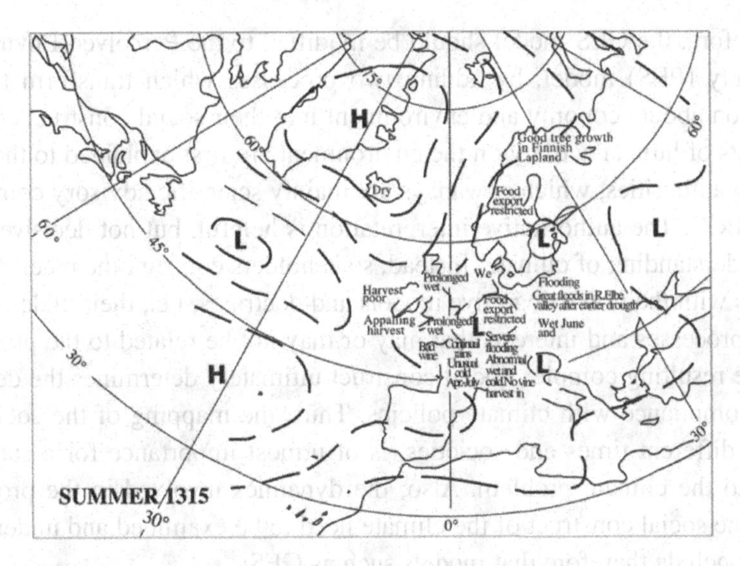

Figure 3: Reconstructed summer mean air-pressure distribution in the year 1315 together with reports about the prevailing weather anomalies (21).

such as the purgatory. Thus, the damage, or adaptation costs, were assessed as being infinite.

Abatement measures considered, were related to a closer adherence to Christian lifestyles. Analogous to the present situation, such an abatement policy was considered as generally benevolent apart from the immediate harvest problems it might remedy. The costs of such a course of action were perceived as considerably smaller than the expected damages. Consistent with such a perspective, the authorities advised their flock "*to atone for their sins and appease the wrath of God by prayer, fasting, alms giving, and other charities*" (22).

Later, climate conditions returned to normal. These developments must have counted as strong evidence to the public and the authorities alike that their climate policy was entirely successful.

Within the context of our contemporary knowledge about climate dynamics, the 1315 case appears to be almost absurd. But we cannot really be confident that our own comprehension of many present environmental crises and their management by society and governments will not appear to future generations equally incongruous. Indeed, what can be learned from this case, is that the GES model is overly simplistic because it implicitly assumes that the costs refer to actual processes. What happens in reality is, however, that the costs are estimated for the perceived processes, and that this understanding is subject to its own dynamics, largely independent of the real processes.

Therefore, the GES model should be modified to the Perceived Environment and Society (PES) model, by adding two processes which transform the hard information about economy and environment into their social constructs (Fig. 4). The effects of human activity on the environment are first explained to the public by certain authorities, which nowadays are mainly scientific advisory committees such as IPCC. The authoritative interpretation is helpful, but not decisive for the public understanding of climate. Instead, stakeholders confront the received interpretations with their own cognitive models and doctrines, i.e., their understanding of many processes and interests that may or may not be related to the problem at hand. The resulting complex social construct ultimately determines the design of and the compliance with climate policies. Thus, the mapping of the social construct, in different times and societies, is of utmost importance for a successful solution to the climate problem. Also, the dynamics involved in the process of forming the social construct of the climate need to be examined and understood.

We conclude therefore that models such as GES:

(1) are informative and useful to discuss the general format of the problem;
(2) lack a crucial module, namely the module that would assist in describing the evolution of social value assignment, or social preferences including conflicts and contradictions in values within and among societies. For a few years, such a figuration of preferences may be taken to be constant but beyond that time scale this process is likely to exhibit significant variations created by social, economic, political and cultural processes.

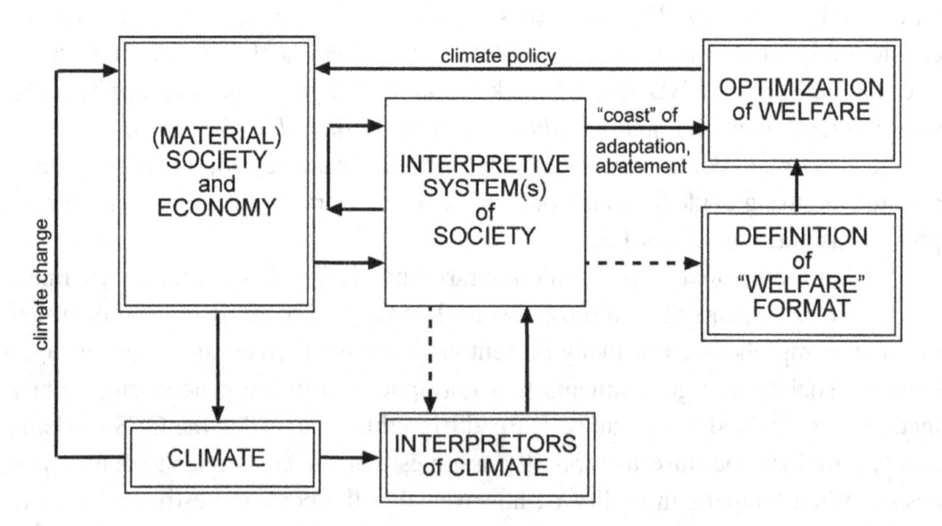

Figure 4: The Perceived Environment and Society (PES) model, which deviates from the GES model in Figure 2 by two additional boxes representing societal processes.

2.4 The social construct of climate

We know of only a few studies that try to describe the prevalent social construct of climate and climate change. In the following, we present some results from an interesting study carried out in the US by Kempton *et al.* (23). This study offers ideas and observations which we consider promising starting points for future research in this direction.

They first interviewed individuals from various social groups in order to identify what lay people think about climate and climate change. Certain ideas were found to be rather widespread, namely that the climate problem is essentially a pollution problem, similar to SO_2 emissions. Thus, an adequate strategy would be to force industry to set up filters. Another frequent (mis)conception was that the emission of CO_2 into the atmosphere would be harmful because it would lead to a depletion of atmospheric oxygen, so that people would suffocate. The natural scientific view of the climate problem was only grasped by a small minority of respondents. On the other hand, preposterous statements such as "I don't know what they're doing up on the moon and shooting those things up there. I think they're disturbing the atmosphere" were voiced by more than one respondent.

These interviews assisted in constructing a questionnaire that was used in a survey of five different groups of respondents, ranging from radical environmentalists to workers who lost their jobs because of environmental legislation. The opinions obtained did not vary much among the different groups. All groups are seriously concerned about the climate problem, and almost all respondents held plain misconceptions about climate, as already found in the initial interviews. For example, 79% of those surveyed agreed with the statement "the weather has become more variable and unpredictable recently" while 43% accepted the possibility of a causal link between changes in weather and the space program.

Obviously, much more research is needed to document what kind of conceptions, and why people have specific conceptions of climate and the climate problem. A relevant line of research in this context would deal with some of the producers of the social construct of climate. In our age, this would certainly include scientists, novelists, journalists, meteorologists, the mass media, and others.

The opinions of climate scientists in the US, Canada and Germany have recently been examined empirically. First results have been published by Bray and von Storch (24, 25). One of the initial conclusions is, "the perception of the risk(s) of global climate change are a product of scientific practice; and the specific hazards variously associated with the event have a close affinity to the scientist's personal belief system." Significant differences by country of residence were

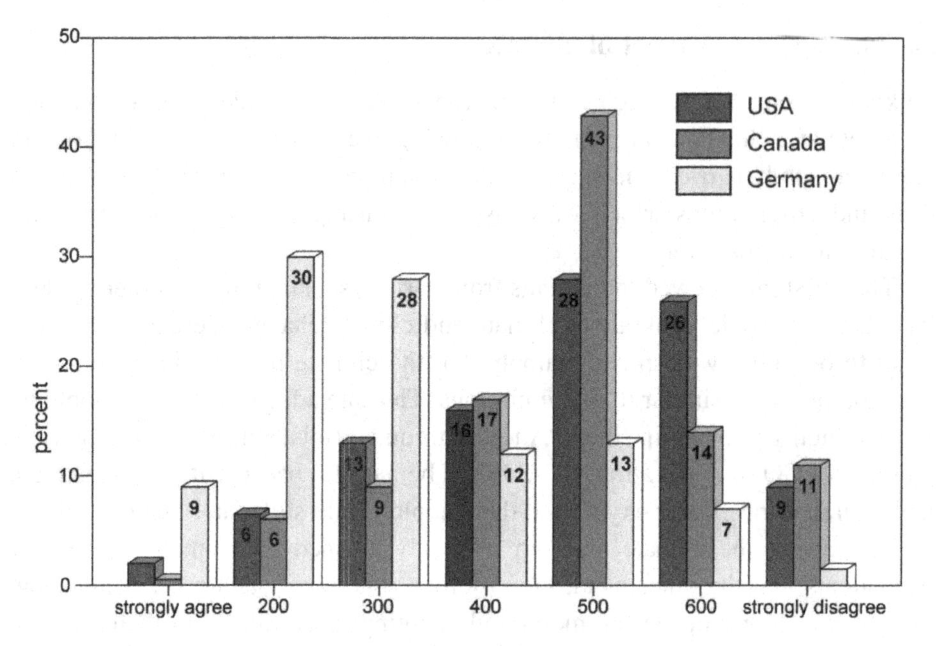

Figure 5: Relative frequencies of answers to the question "scientists are well attuned to the sensitivity of human social systems to climate impacts" from US, Canadian and German scientists, who were asked to answer on a scale from "strongly agree" to "strongly disagree" (24, 25).

found, as exemplified in Fig. 5, in which the answers from about 200 North American and German scientists to the statement "Scientists are well attuned to the sensitivity of human social systems to climate impacts" are summarized. The Germans display greater optimism than their US counterparts while the relative pessimism of Canadian scientists must be related to cultural factors that need to be explored.

We conclude that much more attention and analysis should be specifically devoted to social processes which help or resist, transformation of scientific knowledge into popular beliefs and mental models, and generally to the nature of the social construct of climate and its impact on shaping climate policy in different societies.

2.5 Summary

The three cases presented in our discussion underscore the need to bring the social sciences into climate research. Social scientists could help in understanding the role of climatic determinism and other popular perceptions, the process of social construction of climate-related knowledge and beliefs, and the dynamics of social preferences. Also, the role of natural scientists, who claim to represent "pure" knowledge, but are controlled by various subjective and social mechanisms should be explored in relation to their bringing climate and climate change to the public and political arena. In future, we will need not only future climatic scenarios, but also scenarios of coping with scientific predictions of climatic change as well as scenarios of coping with actual climate change.

Specifically, the following aspects should be studied:

1. What has happened to the doctrine of climate determinism and what climatic events influence under what conditions societies? How far has and can society emancipate itself from climatic conditions? What are the fundamental errors made by Huntington and others?
2. Do the discussions from the last century about natural and/or anthropogenic climate change represent a useful analog for the understanding of the present debate and the present decision process on national and international levels?
3. How can we incorporate the dynamics of social value assignment to transform a GES model into a more realistic PES model?
4. What is the nature of the contemporary social construct of climate and climate change in a comparative perspective, and what changes has this construct undergone in the past years?
5. What is the role of climate scientists in the process of forming the social construct of climate and climatic change?
6. What is the role of other social agents — media, religion, education, the state, etc. — in the formation of the social construct of climate and climatic change?
7. How do we successfully combine social and natural science discourse in the area of climate research?

References and notes

1. Hegerl, G.C., von Storch, H., Hasselmann, K., Santer, B.D., Cubasch, U. and Jones, P.D. 1996. Detecting anthropogenic climate change with an optimal fingerprint method. J. Clim. 9, 2281–2306.
2. Houghton, J.T., Meira Filho, L.G., Callander, B.A., Harris, N., Kattenberg, A. and Maskell, K. (eds.). 1996. Climate Change 1995. The Science of Climate Change. Cambridge University Press, 572 p.
3. Bengtsson, L. 1997. A numerical simulation of anthropogenic climate change. Ambio 26, 58–65.

4. Huntington, E. 1925. Civilization and Climate. Yale University Press, 2nd edition.
5. Wolterek, H. 1938. Klima-Wetter-Mensch. Quelle & Meyer, Leipzig (in German).
6. A more in depth-discussion is offered by Stehr, N. 1996. The ubiquity of nature: Climate and culture. J. Hist. Behav. Sci. 32, 151–159.
7. Beck, R.A. 1993. Climate, liberalism and intolerance. Weather 48, 63–64. The joumal "Weather" is published by the Royal Meteorological Society in London, UK.
8. Pennebaker, J.W., Rime, B. and Blankenship, V.E. 1996. Stereotypes of emotional expressiveness of northerners and southerners: A cross-cultural test of Montesquieu's hypothesis. J. Pers. Soc. Psych. 70, 372–380.
9. Williamson, H. 1770. An attempt to account for the change of climate, which has been observed in the Middle Colonies in North America. Trans. Am. Phil. Soc. 1, 272.
10. For a summary, see Brückner, E. 1890. Klimaschwankungen seit 1700 nebst Bemerkungen über die Klimaschwankungen der Diluvialzeit. Geographische Abhandlungen herausgegeben von Prof. Dr. Albrecht Penck in Wien. Wien and Olmütz, E.D. Hölzel, 325 p. (in German).
11. For a review of Bruckner's work, see Stehr, N., von Storch, H. and Flügel, M. 1996. The 19th century discussion of climate variability and climate change: Analogies for present day debate? World Res. Rev. 7, 589–604.
12. Critical analyses are offered by Sorokin, P. 1928. Contemporary Sociological Theories. Harper & Row Publishers, New York (Reprint 1956), 783 p.
13. A recent critical analysis is given by Nordhaus, W.D. 1994. The ghosts of climate past and the specters of climate future. In: Integrative Assessment of Mitigation, Impact and Adaptation to Climate Change. Nakicenovic, N., Nordhaus, W.D., Richels, R. and Toth, F.L. (eds). IIASA, May 1994, pp. 35–62.
14. Harding, G. 1968. The tragedy of the commons. Science 162, 1243–1248.
15. Nordhaus, W.D. 1991. To slow or not to slow: The economy of the greenhouse effect. Econ. J. 101, 920–937.
16. Hasselmann, K. 1990. How well can we predict the climate crisis? In: Environmental Scarcity — the International Dimension. Siebert, H. (ed.). JCB Mohr, Tubingen, pp. 165–183.
17. The case of a transient evolution of a highly idealized system has been worked out by Tahvonen, O., von Storch, H. and von Storch, J. 1994. Economic efficiency of CO_2-reduction programs. Clim. Res. 4, 127–141.
18. Hasselmann, K., Hasselmann, S., Giering, R., Ocaña, V. and von Storch, H. 1997. Sensitivity study of optimal CO_2 emission paths using a simplified structural integrated assessment model (SIAM). Clim. Change 37, 345–386.
19. If the countries cannot agree on a joint policy, the problem may be cast into the format of differential game theory, see Hasselmann, K. and Hasselmann, S. 1996. Multi-actor optimization of greenhouse gas emission paths using coupled integrated climate response and economic models. Proceedings, Potsdam Symposium. Earth System Analysis. Integrative Science for Sustainability, 1994.

20. Stehr, N. and von Storch, H. 1995. The social construct of climate and climate change. Clim. Res. 5, 99–105.
21. Lamb, H. 1987. What can historical records tell us about the breakdown of the medieval warm climate in Europe in the fourteenth and fifteenth centuries — an experiment. Beitr. Phys. Atmos. 60, 131–143.
22. Kershaw, I. 1973. The great famine and agrarian crisis in England, 1315-1322. Past Present 59, 3–50.
23. Kempton, W., Boster, J.S. and Hartley, J.A. 1995. Environmental Values in American Culture. MIT Press, Cambridge, MA and London, 320 p.
24. Bray, D. and von Storch, H. 1996. Inside science — a preliminary investigation of the case of global warming. MPI Report 195, 58 p.
25. Bray, D. and von Storch, H. 1996. The climate change issue. Perspectives and interpretations. Proc. 14th Intl. Conf. Biometeor, 1–8 Sept. 1996, Lubljana, Slovenia. See also Auer, I., Bohn, R. and Steinacker, R. 1996. An opinion toll among climatologists about climate change topics. Meteor. Z. NF 5, 145–155.
26. This essay is the result of an extensive cooperation between the two authors, one of which is a social scientist and the other a natural scientist. This cooperation was made possible by several visits of Nico Stehr to the Max-Planck-Institut für Meteorologie in Hamburg and the Institute fur Gewässerphysik in Geesthacht. We are indebted to the Max-Planck-Gesellschaft, the GKSS Forschungszentrum and the Thyssen Foundation for their generous support of this cooperation. The comments of the two anonymous reviewers were most helpful. Hans von Storch thanks Klaus Hasselmann and Dennis Bray for many stimulating discussions.

3. The Climate in the Public Mind[‡]

Our contribution deals with a distinction in the perception and handling of climate, as it can be observed in modern societies. This is the distinction between the everyday perception and the scientific perception of climate and weather. Everyday impressions and beliefs about climate — for example, about its power to influence the conditions of human life, the development processes of human societies, but also the differences between people, such as their economic success, their health, or their well-being — go back much further than the scientific ideas about climate and weather.

The scientific views are barely a century old. Observations of climate, which can, for example, rely on systematic measurement methods, only began in the late 19th century. However, climate science has not yet succeeded in replacing the everyday understanding of climate and weather. Therefore, we are dealing with what can be called the societal construct of climate on the one hand, and the scientific construct of climate on the other hand. The convergence or contradictions that can be identified between these constructs have a not inconsiderable significance for practical climate policy. They also affect the efforts of climate scientists to make their results and practical conclusions comprehensible to the public. In these efforts, climate scientists always encounter the social construct of climate, which strongly influences people's everyday perceptions and affects the success of science's communication with the public, the media, and politics.

The divergence of everyday and scientific beliefs has only gradually emerged. Only a few decades ago, many scientific papers contained observations and conclusions about the influence of the climate — keyword here is climate determinism — which scientifically underpinned the everyday awareness of the power of the climate. The differentiation between societal and scientific constructs is more recent. However, it is not uncommon to observe that climate scientists still promote or advocate climate determinism today — keyword here would be climate catastrophe.

There are also remarkable similarities in the everyday and scientific conceptions of climate and weather. These include, for example, the certainty with which people on both sides are convinced of their own conceptions of climate — on the one hand, concerning the statements about global climate change provided by scientists' climate models, and on the other hand, with regard to the certainty with which people in everyday life speak of the overwhelming power and the unique influence of climate on human activity.

[‡]This section originally appeared in von Storch, H. and N. Stehr: "Das Klima in den Köpfen der Menschen," pp. 280–291 in Walter Hauser (ed.), *Klima*. Das Experiment mit dem Planeten Erde. München: Deutsches Museum, 2002. Translated from German.

Thomas Bernhard (1986, pp. 18–20) — who certainly did not become famous by the type of compliments he made about his Austrian compatriots — refers in his book 'Der Untergeher' exemplarily to this constellation of climate influence and character when he spitefully states: "The people of Salzburg have always been as awful as their climate, and when I come to this city today, not only is my judgment confirmed, but everything is even more awful. ... The climate of the foothills of the Alps makes people sick of mind, who fall prey to stupor at a very early age and who become mean spirited with time. ... This climate and these walls kill sensitivity."

The conviction, shared to this day by many a climatologist, that climatic conditions are a key, if not the key, to understanding the peculiarities of the inhabitants of different continents, regions, and places, can be observed in many cultures. The Japanese ambassador Kume Kunitake (1878), for example, describes his experiences with Germany and Austria: "The character of the Germans is by nature thorough and deliberate, and therefore they lack perspicacity and dynamic in undertakings. When, however, exact and careful work is needed, one encounters an astonishing perseverance which must be mentioned with praise. The Prussians live in harsh and cold areas in the north. This poverty, however, did not discourage them, but increased their perseverance by it ... it seemed to me that the atmosphere in Berlin therefore radiated a certain arrogance and aggressiveness. Austria, on the other hand, is a country blessed with fertile soil and a mild climate. Culture has long flourished there in its famous cities. This wealth favored the emergence of a softer nature of Austrians. They like to indulge in the urban lifestyle and its refinements, and rave about culture and art. Splendor and lavishness go hand in hand."

However, the comparison of Thomas Bernhard's observations with those of the Japanese ambassador, demonstrates that the characteristics of the people and their societies, which are supposedly a consequence of their special climatic conditions, turn out to be quite arbitrary and contradictory. The reference to the responsibility of the climate only disguises deeply rooted social prejudice.

The importance and the persuasiveness of the social construct "climate" is based on the fact that this phenomenon, as far as we know, has always interested people intensively — climate not in the sense of an "average weather," because that does not exist, but in the sense of the variations of the weather or, more precisely, the statistic of the weather. Whereby one has to distinguish between the "normal fluctuations" and the extreme events, which may occur, but are statistically seen rare; indeed, so rare that they are not considered part of the culture of everyday life.

Climate is an environmental condition that affects and influences our lifestylcs. However, today this influence is mainly due to our climate perceptions and less so to climatic events. In everyday life, we have disconnected ourselves from the weather in many regions of the world. For a large part of the day, of the months and of the seasons, we live independent of local weather conditions. In former times, i.e., until 100 years ago, the influence of the climate on human living and working conditions was much more direct, as the climatologist Eduard Brückner (1863–1927) already proved in his dissertation in 1890: Agricultural production and thus the well-being of an agricultural society was subject to suitable climatic conditions. Thus, communication and the movement of people's goods were hindered or prevented when rivers and harbors froze over and shipping had to be stopped. Today, however, trains run and airplanes fly. The share of agriculture in the gross national product of the industrial nations and the knowledge societies on the horizon is small. At the same time, technical discoveries and developments, such as the mass production of air-conditioning systems, make it possible for people in all climatic regions of the world to live a life that is hardly dependent on the respective climatic conditions. Nevertheless, the climate still influences our lifestyle today, albeit in a different way — concern about man-made climate change has led to calls for climate protection: For a reorganization of society and the economy in a climate-friendly way, since the climate is "tipping over" and thus the celebrated emancipation from the environment is threatening to turn into its opposite.

Climate and weather are experienced by everyone, and these experiences form our attitudes towards climate. Especially extreme events determine our understanding of climate. Paradoxically, however, extraordinary weather phenomena strengthen our confidence in the power of the normality of the climate, just as the deviant behavior of fellow human beings reminds us of the existence of social values and norms to which we should adhere. Extreme events make us aware of the reliability of the climate. They represent a kind of reassurance. Since a return to normality is expected once the crazy weather event subsides, such events strengthen, in the end, our confidence in the reliability of the climate, for example, in the reliable sequence of seasons or in the fact that a cold period will be followed by more pleasant temperatures and that heavy storms will be replaced by mild winds.

Climate is a universal subject of discourse, explanation, and concern. Everyday knowledge about climate is ancient. It is practical knowledge — it explains things in a useful and understandable way without necessarily being accurate in the sense of modern climate impact research. It is ubiquitous. It is socially constructed knowledge that is refreshed over time by scientific knowledge which may become outdated soon. It is effective knowledge, not least because it influences public opinion, the shaping of climate policy and the thinking of climate (impact) researchers. It competes with contemporary scientific knowledge about climate and climate impacts. It is the climate in people's minds.

What are the typical characteristics of mundane climate awareness? There are two main aspects, to which we have already briefly referred. First: That climate has a strong influence on the human body (think of the widespread thesis that the skin color of people is the result of different climates or that the reproductive power is decisively determined by the climate), and on psychological and social qualities; it determines our well-being or is responsible for our failures. Secondly: The climate is getting worse and worse. Responsible for this deterioration — which, by the way, was not discovered or invented by climate science, because climate science was convinced until a few decades ago that climate in historical times was primarily a static phenomenon — is human action, which is reflected in a change of weather and climate.

We will first outline both aspects of climate consciousness in more detail under the headings of "climate determinism" and "climate catastrophe," before pursuing the question of why these are not merely irrelevant details of the history of ideas, but views that still play a significant social role today. Finally, we will recommend a science-policy response aimed both at containing the influence of pre-scientific knowledge and at showing the contingency of contemporary scientific knowledge.

3.1 Keyword: Climate determinism

Already the classical Greek philosophers held the thesis that the climate was responsible for the fact that people on different continents, regions and places of the world differed; some areas were favored — as a rule always their own —, others were disadvantaged by the climate — as a rule the residential areas of the barbarians, the strangers, the others. This idea was taken up again later, in the age of enlightenment, by Montesquieu, Herder and other thinkers. Later, this conviction of the special power of climate, which was accepted as obvious, became the subject of learned scientific discourse. The scientification of these everyday ideas took place on the basis of comprehensive, supposedly quantitative-objective methods. For example, the leading American geographer Ellsworth Huntington (1876–1947) examined performance characteristics of students and workers in the U.S. and designated certain temperature ranges and characteristics of the seasonal and weather variations as optimal. The weather should be not too cold and not too hot, and not too monotonous. Huntington concluded emphatically from his empirical observations that the correlations and dependencies he derived from short-term changes in weather patterns should apply to entire populations and should for instance be decisive in the choice of locations for production sites and political institutions (such as the headquarters of the United Nations).

Because a colonist or tourist in the tropics tires quickly under the hot and humid weather conditions, moves more slowly, and has fewer children when he settles there, all the inhabitants of the hot climates are listless, unambitious and unproductive — and permanently. If one looks at encyclopedias of the time of the outgoing 19th century, this view of climate and climate effect is definitive. Under the keyword "climate," one regularly finds the reference that culture and civilization can blossom only in the temperate maritime climate of the middle latitudes. The inhabitants of the hot zones are quickly aging, overactive, irrational, and tyrannical. The inhabitants of subpolar zones, on the other hand, are fat, immobile and childlike. Note — it is not the merit of humans in the maritime climate of the middle latitudes to have developed culture and civilization; the obligation with spreading the culture arises as a must, to colonize the world and to make it content with Christianity, hamburgers, refrigerators and other blessings of civilization.

The influential physician and social psychologist Willy Hellpach (1877–1955) held this view of the comprehensive, fateful consequences of climate with prodigious conviction and resonance just a few decades ago: "In the northern part of a region of the earth the traits of sobriety, astringency, coolness, composure, the willingness to make an effort, patience, tenacity, rigor, the consistent use of the intellect and will predominate — in the southern part the traits of liveliness, excitability, impulsiveness, the sphere of feeling and fantasy, the more sedate letting-go or instantaneous flaring up. Within a nation, its northern populations are more practical, more reliable, but more inaccessible, its southern ones more musical, more accessible (jovial, amiable, talkative), but more unsteady."

Accordingly, collectivities and individuals are chained to their climate; they make what is possible given the constraints of the climatic conditions. But despair to her who leaves her ancestral climate. The climate accompanies the migrant. If the "new" climate corresponds to the original, in each case optimal, then it is good. Therefore Scandinavians, so again Huntington, could live very well in today's federal state Washington in the northwest of the USA. But if the climate encoded into the genes of a human being is not in harmony with that of his place of residence, then serious problems arise. If, for example, the "white" man settles in the tropics, he perishes. We encounter this unhappy being in Humphrey Bogart's movies. That is why the Africans should rather stay in Africa or the blacks in the southern states of the USA. For their own sake; because who leaves his ancestral climatic region, becomes not only unhappy but also unproductive. Only the Chinese are out of line — according to a judgment by Willy Hellpach, they are "climate insensitive." In view of the even for climate determinists' undeniable fact that Chinese are "successful" in all climate regions of this world, the notion of determinism is rescued by the auxiliary construction of an alleged climate insensitivity of the Chinese people. Why the rest of mankind is sensitive, however, Hellpach fails to make clear. Climate determinism promotes, reinforces, and conceals social prejudices.

Is climate determinism today simply an outdated, curious detail of the history of ideas? Hardly. It used to be widely accepted science-based knowledge; in the field of geography, climate was a standard element for explaining differences between societies, as was the reference to the nonuniform distribution of natural resources (mineral resources) or other geographic determinants. However, the explanatory factor "climate" had to compete with other scientifically or pre-scientifically accepted influences on societal development — especially with the factor heredity. One could not escape its "innate" climate any more than one could escape other innate characteristics, such as race. After the shipwreck of racism, climatic determinism also disappeared from the stage of science. But in everyday life in many societies, both are still firmly rooted. To be sure, at present one rarely finds evidence of an infectious climate determinism in public or published opinion. But everyone seems to know that the Swedes are so much more efficient than Nigerians because the latter live in a comfortable climate and are therefore carefree and listless, while the former are constantly plagued by their weather extremes and therefore think up clever things to make their lives more independent of the weather and generally more efficient and beautiful.

At present, climate determinism is returning "clandestinely," so to speak, in scientific garb. And this in a certain branch of climate impact research. It has been and still is a widespread practice to estimate the consequences and risks of expected anthropogenic climate change by designing models of a future state of society in which climate change is the only dynamic element, while other societal conditions — i.e., economy, politics, social consciousness, international relations or technical-scientific development — are treated as constant factors. This is true, for example, of a much-acclaimed study on the future state of the forests in Brandenburg, Germany. However, the fact that social institutions, economic needs, values, and technologies are bound to change in the future and will presumably be much more responsible for observable ecological and societal changes than the assumed climate changes is often deliberately overlooked. It is the historical climate determinism that influences such studies and what can be regarded as probable future events with the consequence that the (undoubtedly uncertain) societal development with its probably still increasing flexibility, it's hardly predictable modernization processes and adaptation attempts to the changing climate is underestimated. In light of such premises, it is not surprising that efforts to protect the climate from society do not place greater emphasis on adaptation strategies, but rather aim almost exclusively at reducing the release of greenhouse gases as an appropriate response to climate change.

3.2 Keyword: Climate catastrophe

We hear it almost daily: The climate is getting worse; it seems it has always been getting worse. Seasons used to be more consistent; there was usually snow at Christmas. The weather is less predictable than before; the weather goes crazy, the storms are getting wilder. Overall, the climate is becoming less favorable. In the 1950s, it was due to the atomic bombs. Already Gustav Gans in a Donald Duck story of those years knew that. Another fifty years earlier, the rifle and gun fire and the transatlantic shortwave radio were responsible. And before that, the lightning rods that aroused the ire of angry citizens, as the Neue Zürcher Zeitung reported in 1816. Before that, witches and sinful life was punished by God with bad weather. After 1970, it was said that air pollution was accelerating the transition to the new ice age, which was imminent anyway. At present, global warming is responsible for spectacular natural events. Humans have always been to blame. One can write a history of societal development as a history of climate catastrophes.

But this time it is different. This time it is serious. Greenhouse gases in the atmosphere are increasing. There is no doubt about it. The long-term CO_2 data from Mauno Loa in Hawaii show it clearly. And scientific climate research is almost unanimous in its assessment that increases in greenhouse gases in the atmosphere are accompanied by climate changes. So far, the shifts have been relatively small, but they can be documented; in Germany, the average temperature has risen by 1 degree in the past hundred years, and in a few decades — even if there is a drastic reduction in greenhouse gases by then — the climate changes will probably be more pronounced. For the time being, we are only talking about slight increases in globally averaged temperature and sea level. It is uncertain whether the changes observed so far are precursors of much more severe deviations from the previous state.

Is the expected climate change a "catastrophe"? Only if it is accompanied by catastrophic consequences. And these are not to be found unless one concentrates on such unlikely events as the disappearance of the Gulf Stream, which was allegedly already cut off once by the Aswan Dam and the resulting lack of freshwater inflow into the Mediterranean. The consequences include heat waves and climatic deaths in metropolitan areas. But U.S. experience shows that it is the social marginality of certain social strata in America's metropolitan areas that causally leads to a greater number of heat deaths under extreme weather conditions. It is not the increased temperatures themselves. Similarly, the risk of malaria and other diseases spreading to regions of the world where these diseases no longer occur is often cited as a dangerous consequence of climate change. Malaria once existed on an endemic scale in England and Holland; it was eradicated through medical innovation, hygienic measures, and social organization. Malaria is largely a disease of the poor. And let's not forget Bangladesh — where, even without the consequences of anthropogenic climate change, people must fear every tropical cyclone because they lack adequate coastal protection.

The expected climate changes will require adaptations, probably difficult adaptations. The reference to victims, damages and risks in present and future extreme situations is always also a reference to social causes, which bring about many conceivable catastrophes in the first place — but thus in principle also make them avoidable. Even without a climate catastrophe, it is worthwhile to use energy and raw material reserves sparingly and sustainably. Adaptation to climate change and reduction of greenhouse gases do not contradict each other as political strategies.

3.3 Practical application?

In March 2002, one of the media stars of German climate research wrote in *Die Zeit*: "The breaking-off of huge ice masses in Antarctica a few days ago has brought the climate problem back into the focus of public attention. While the causes of this spectacular event are not yet clear, one thing is certain: Global climate change is in full swing." That's a remarkable formulation, insofar as the first sentence is true. The breaking-off of icebergs in Antarctica or the storms in Germany and other European countries in the summer of 2002 generated extensive media coverage, as is always the case with extreme events. There was a lot of public palaver about climatic reasons, especially about the consequences of the greenhouse effect. The experts from the *Alfred Wegener Institute for Polar and Marine Research* have contradicted this, but still! The second sentence is also true — the causes are not clear. But then follows an astonishing twist: The storms and the breaking-off of the Antarctic ice are conclusive signals, if not adequate evidence, of climate change. Apparently, there are two kinds of knowledge in the researcher's mind — the scientific one of the first two sentences and another, pre-scientific knowledge that directly links the melting of polar ice caps to frightening consequences for global water levels.

The "climate in the public mind" is also in researchers' minds. We have criticized this attribution of risks, damages, deaths, destructions as direct consequences of climate change. The fact that deaths and injuries, immeasurable economic damage or health risks are understood as an expression of natural disasters may be in the interest of those who bear the responsibility for having tolerated or even promoted the fact that certain people have put themselves or had to put themselves in danger. But the real cause of the victims and the damage to be lamented are political and other failures. For an extreme weather event to become a catastrophe, it must pass through the bottleneck of society and its conditions.

In fact, it is probably the convergence of pre-scientific and scientific ideas about climate that allows some scientists to be extraordinarily successful in the media: They often just confirm what everyone already knows and feels. But exactly these commonalities and the question why natural catastrophes are not necessarily natural catastrophes but, for example, manifest a failure of society towards certain sections of the population, should be part of an interdisciplinary research initiative that does not exist so far.

We need the duality of social (or cultural) and natural sciences for better understanding what is the traditional "climate in our heads" — and what is scientifically tenable knowledge. So that we can help natural scientists separate themselves from pre-scientific, unconscious concepts and bring lay people up to date with the latest scientific knowledge. This will allow us to understand what it means that many terms are used with dual meaning, as well-defined technical terms and as lay expressions with a large yard of connotations. We need a mapping of the cultural and social constructs of climate, a history of climate, climate perception, and climate research. On this occasion we will be astonished to find that already at the end of the 19th-century parliamentary commissions in Prussia, Russia and Italy were thinking about anthropogenic climate change and the possibilities of its prevention. At that time, climate change was not global but regional, and the cause was not greenhouse gases but deforestation. But otherwise, it was very similar to today. At that time, of course, it was a false alarm, but today it is not, as the majority of climate researchers states.

Literature

Bernhard, Thomas, *Der Untergeher*, Frankfurt am Main: Suhrkamp 1986.

Huntington, Ellsworth, *Civilization and Climate*. Third Edition with Many New Chapters, New Haven, Conn.: Yale University Press 1990.

Stehr, Nico, »Trust and climate«, in: *Climate Research* 8 (1997), S. 163–169.

Stehr, Nico, und Hans von Storch, *Klima — Wetter — Mensch*, München: C. H. Beck 1999.

Stehr, Nico, und Hans von Storch, »The social construct of climate and climate change«, in: *Climate Research* 5 (1995), S. 99–105.

4. Climate Research and Policy Advice: Scientific and Cultural Constructions of Knowledge[§]

Abstract

The section is a call to cultural sciences for helping climate science to establish a sustainable practice of policy advice concerning man-made climate change. As a climate scientist engaged in communication with stakeholders and the media, mostly in Germany, the author has noticed a notable discrepancy between scientific knowledge about climate change, and the understanding in the public at large, specifically as fostered by the media and some publicly visible climate scientists. In this essay, this discrepancy is analyzed to some extent and framed as the presence of two competing types of knowledge, namely a body of knowledge named "scientific construct" and another body of knowledge named "cultural construct" of man-made climate change. The relationship and the dynamics of these two knowledge claims are not well researched. In order to understand the dynamical interaction of the different knowledge claims significant efforts from cultural sciences are needed. Unfortunately, so far these disciplines do not often consider this field. Two examples of useful analyses are presented as examples.

4.1 Knowledge about climate change

Science has established that processes of human origin are influencing the climate — human beings are changing the global climate. Climate is the statistics of the weather. In almost all localities, at present and in the foreseeable future, the frequency distributions of the temperature continue to shift to higher values; sea level is rising; amounts of rainfall are changing. Some extremes such as heavy rainfall events will change. The driving force behind these alterations is above all the emission of greenhouse gases, in particular carbon dioxide and methane, into the atmosphere, where they interfere with the radiative balance of the Earth system.

This is the scientific construct of human-made climate change. It is widely supported within the relevant scientific communities, and has been comprehensively

[§]This section originally appeared in von Storch, H.: "Climate research and policy advice: Scientific and cultural constructions of knowledge," *Env. Sci. Pol.* 12:741–747, 2009.

formulated particularly thanks to the collective and consensual efforts of the UNO climate council, the "IPCC."[2]

Of course, there is no complete consensus in the scientific community, so that speaking of "the scientific construct" is a simplification, which is applied in this essay for describing the contrast to the cultural construct, which is equally not just one construct but features many different variants. What is stated in the previous paragraph is the core of the consensus, and may therefore represent the core of "the" scientific construct.

What else do people know about climate and climate change? That the climate really is changing because of human activity — due to deforestation, for example, as well. That the weather is less reliable than it was before, the seasons more irregular, the storms more violent. Weather extremes are taking on catastrophic and previously unknown forms.

The cause? Human greed and stupidity. The mechanism: The justice, or the revenge, of a nature that is striking back. For large chunks of the population, at least in Central and Northern Europe, the mechanism is obvious.[3] In old times, the adverse climatic developments were the just response to a God angered by human sins; this approach is also today, in particular in the US and possibly in the UK, sometimes invoked. An example is provided by the former Chair of the IPCC, Sir John Houghton, who expressed his conviction that God would speak to the public through disasters (Welch, 1995). Or, as it is put on the back cover of an alarmistic book from an alarmistic book in 1989 by Milne, *Our Drowning World* (Milne, 1989): "... we shall be engulfed by the consequences of our greed and stupidity. Nearly two-thirds of our world could disappear under polar ice cap water. ... For this will be the inevitable outcome of industrialization, urbanization, overpopulation and the accompanying pollution." An enlightened variant is suggested by Lovelock in the framework of his Gaia hypothesis, when he speaks about *The Revenge of Gaia: Why the Earth is Fighting Back – and How We Can Still Save Humanity* (Lovelock, 2006).

Finally, there was also the idea of "climatic determinism" (see Stehr and von Storch, 1999), which can be traced back to classical Greek literature and was

---

[2] Just to avoid misunderstanding — I am a natural scientist, have contributed to the IPCC process and consider the "scientific construct" realistic. I do not ascribe to the cultural construct as a useful and realistic concept for understanding climate, climate change and climate impact, which I try to describe in the following paragraphs. The cultural construct is, however, a powerful concept, which deserves scientific analysis and attention.

[3] On 14 August 2002, the reputable Swedish daily newspaper *Dagens Nyheter* wrote "Naturen slår tilbaka våldsamt." (Nature strikes back violently), when reporting about a disastrous flooding in the Czech Republic.

widely accepted in western thinking in the 18th century up to the middle of the 20th century. Some elements of this school of thought continue to have an effect upon our western culture. One of these elements is the understanding that human beings have to live in balance with that climate that is suitable for them. If this climate changes, civilization is at risk; whole cultures perish on such an occasion — for instance, Native American cultures in North America, the Viking settlements in Greenland. No wonder, then, that in German usage, the term is *Klimakatastrophe*, "climate catastrophe," and not "climate change."

This is the *cultural* construct of climate change, particularly in the German-speaking countries, but similarly, and widely, in other areas of the West as well. It has to be stressed, again, that referring to just one well-defined construct is a simplification; obviously, the cultural construct takes many different forms, displays various nuances — but what is described above represents something like a standard core of such statements. Obviously, the scientific construct is hardly consistent with this cultural construct.

These two constructs, scientific and cultural, are competitors in the interpretation of a complex environment; two protagonists on the market of knowledge. Their inconsistency makes them incompatible, but sometimes the two forms are nevertheless mended, the efficacy of the modernized construct so formed increases; its scientific basis, however, becomes narrower. Public acceptance rises; its robustness in the face of scientifically verifiable facts sinks; but its utility for pursuing certain value-driven political agendas is improved; at the same time, science is corrupted. Of course, scientific practice — and thus its construction of explanations and theories — are influenced by the cultural construct in any case; because, after all, we simply cannot live free of our culture. Our culture conditions us in our point of view, steers us in our formulation of questions, in our willingness to regard answers as sufficient to build an argument. Needless to emphasize that there is no pure scientific construction but at least a kind of approximation of such.

4.2 The arena of public attention

At some point in the 1960s or 1970s, the idea of *Bringschuld of science* was created in Germany — when related to science this term "Bringschuld" denotes the ethical obligation of scientists to pay their debt to society by informing it about existing, arising and possible future dangers. In the past, science had all too often turned a blind eye to such dangers, and instead made itself the willing lackey of such scientific and technological developments as eugenics and nuclear power. Rewarded with financial incentives, recognition, and the satisfaction of a sometimes perverse curiosity, science looked on without lifting a finger and without taking responsibility. It was time to put an end to such inaction, the argument was. Scientists were to see their activities in a social context; to inform the public on their own initiative, without waiting to be asked — so that the public could then decide, democratically, what made sense and what did not.

What do we scientists — who may well be experts in our field, but who are otherwise laypeople like anyone else — know about the dangers at hand? It often happens that the perceived dangers lie outside the realm of our expertise; that is, the we "experts" operate with culturally constructed knowledge — not, however, as the onlooking public believes, with scientifically constructed knowledge. Thus it is not the best knowledge that is put into action, but rather knowledge claims. Pretensions to interpretation, and to power, disguised as science.

It is not that there are few dangers, real or perceived; rather, there are many such dangers. They vie with one another in competition for public attention. The public, including us scientists, is only able to "process" a limited number of topics at length; just how many is unclear, but it seems implausible to me that there would be at the same time more than ten or maybe twenty. Some of these topics are "givens," such as the national soccer league, for instance. How are the topics that come to the fore chosen? One would hope that the deciding factor might be the urgency of the topic, the social or economic relevance, but that is most certainly not the case. Perhaps it is their entertainment value, even their fear-mongering value, the challenge they pose, or even, in terms of the interpretive order, their assurance that all is well with the cultural construct.

But of course one can attempt to "push" one's own field of expertise into the arena of public attention. The attributes needed to accomplish this must then be added — by means of exaggeration, perhaps; by implicit associations; by exploiting the cultural construct, that is to say, what the public in any case recognizes as being "correct." "*Waldsterben*,"[4] as a result of pollution, was certainly one such topic.

Whether fulfilling the terms of *Bringschuld* is beneficial or harmful for the individual certainly depends upon the social context. When everyone was enthusiastic about scientific and technological progress — when the German weekly "Micky Maus" comics,[5] in the series "Our Friend, the Atom," described a golden

---

[4] "Forest dieback," which was particular in the 1970s and 1980s as an icon of human destruction of the environment. It was most popular in Germany, and the German term "Waldsterben" was introduced into the French and English language.

[5] This weekly journal addressed children and juveniles with Walt Disney's comics since 1951 in Germany. Facing opposition because of the comic-character, which was considered inferior to "real" literature, the editor added an supposedly informative article-style mid-part on social and technological issues.

future of ubiquitous nuclear energy to their young readers — then no one paid any heed to the possible drawbacks of these advances. Today, however, with a skeptical attitude toward scientific and technological progress, particularly when it seems to be documented in our immediate environment in the form of masts, noises or smells, a scientifically presented assessment of the dangers is appreciated by society, particularly when such an assessment confirms prior knowledge and thus is recognized *a priori* as correct in any case. This appreciation can take a multitude of forms: A career, public attention and recognition, better working conditions, personal satisfaction in the belief that one has made the world a better place.

Satisfying the demands of *Bringschuld* is no longer an altruistic act nowadays, but rather a productive element in a marketing strategy. *Bringschuld* of science has led to a massive influx of proclaimed dangers into the arena of public attention. Environmental science, and not only environmental science, has become "postnormal."

4.3 Postnormal science between the requirements of policy and the media

The quality of being "postnormal" was introduced into the analysis of science by the philosophers Silvio Funtovitz and Jerry Ravetz (1985). In a situation where science cannot make concrete statements with high certainty, and in which the evidence of science is of considerable practical significance for formulating policies and decisions, then this science is impelled less and less by the pure "curiosity" that idealistic views glorify as the innermost driving force of science, and increasingly by the usefulness of the possible evidence for just such formulations of decisions and policy. It is no longer being scientific that is of central importance, nor the methodical quality, nor Popper's dictum of falsification, nor Fleck's idea of repairing outmoded systems of explanation (Fleck, 1980); instead, it is utility that carries the day. The saying "Nothing is as practical as a good theory," attributed to Kurt Lewin, refers to the ability to facilitate decisions and guide actions. Not correctness, nor objective falsifiability, occupies the foreground, but rather social acceptance.

In its postnormal phase, science thus lives on its claims, on its staging in the media, on its congruity with cultural constructions. These knowledge claims are raised not only by established scientists but also by others, self-appointed experts, who frequently enough are bound to special interests, be they Exxon or Greenpeace.

Currently, climate research is postnormal (Bray and von Storch, 1999). The inherent uncertainties are enormous, since projections of the future are required, or rather: of futures — such futures that can only be represented using models, where conditions will prevail that no one has yet observed. We simply do not know exactly how the cloud cover will alter if temperatures and water vapor content change, or which will win the upper hand in terms of the balance of the Antarctic ice mass — increased precipitation in the heights or melt-off at the edges. Our knowledge is inadequate not because the scientists are incapable, but rather due to the meager facts available, the incomplete data, which moreover span too short a time period. Certainly, there are arguments that point to one answer or the other, and considerations of plausibility allow us to exclude certain developments as unlikely or even impossible. There remains, however, a residual uncertainty that will possibly never be resolved or will be reduced only in the course of years, or even decades. In this situation, the representatives of social interests seek out those knowledge claims that best support their own position. One need only recall the *Stern* report [see the critique by Pielke (2007) or Yohe and Tol (2008)], or the regular press releases of US Senator Inhofe. Not only are those knowledge claims that seem suitable picked out and placed into a matching overall picture; however, new and idiosyncratic knowledge claims are also constructed, so that in the end, a bizarre accumulation of claims is produced — claims that sometimes seem arbitrary, such as that the number of patients with kidney stones will increase (Brikowski *et al.*, 2008), for instance, as a result of human-made global warming. The scientifically untenable film "The Day after Tomorrow" is praised by high-profile scientists as an aid to awareness-raising; political and scientific achievements are intermingled by awarding the Nobel Peace Prize simultaneously to Al Gore and the IPCC; politicians disguised as professors pronounce to the public necessary reactions to climate change. Along with these alarmist tendencies, there is also the skeptical counterpart, represented in such grossly misleading products as "State of Fear," by the otherwise admirable Michael Crichton, or the film "The Great Swindle." All of this is typical for a postnormal science.

In the daily course of events, there are many opportunities for both the individual and powerful scientific organizations to draw public attention to themselves. But there remains not only among natural scientists a gnawing sense that this practice simply cannot be that which we describe as more or less "good science," where it is the argument, the critical inquiry, the well-constructed test, the unconventional idea that lies outside the prevailing paradigm, that effects progress, and not science's usefulness for putting through a policy perceived or described as correct.

What appears in *Science* and *Nature* is often prematurely published research; it stimulates the educated readership's imagination and sometimes its fears — and after a few years, it often proves to be in need of revision anyway. But this revision is ultimately the mechanism that extricates science from the whirling eddy of post-normality. When the caravan of public attention turns to other topics, then normal science takes hold again and compromises with the required usefulness, the *Zeitgeist* and political correctness can be revised. On a smaller scale, we can already see this revision of details (but not the overall assessment) in the field of climate research: For instance, in the case of the so-called "hockey stick," the premature closure of debate on the question of historical temperature fluctuations; or of the perception, pushed by some re-insurance companies, that the risk of storms has increased; or the perspectives that anthropogenic climate change would be associated with the breakdown of the Gulf Stream or the disintegration of the West-Antarctic Ice Shield within decades. All popular claims for a while among media and activists, but oversold and not critically revised.

4.4 A role for the social and cultural sciences

For us, as scientists involved in this matter, the question is: How do we deal, here and now, with this postnormal situation; for we accept both demands — good science and good advice for the public — as justified. The solution can actually only be this: That we do what we do best, at least in principle, namely analyze the situation scientifically. But we natural scientists can do this only to a limited extent. We already suspect that the process of science is a social process; that we are not always quite objective, at least when we frame questions and accept explanations; that we are conditioned by our different cultures. That the advance of individuals into important positions often has less to do with science, and rather more to do with social and political acceptance.

In order to give our analysis depth and substance, we need the skills of the social and cultural sciences. My personal experience, which is admittedly limited, informs me that up to now, however, these sciences have largely kept their distance. What I have heard are occasional and general hints that everything would be socially constructed and relative — which I consider mostly signs of an unfortunate refusal to go into concrete detail, which would be unavoidable for any real synergy. It is annoying when colleagues from these fields obviously fail to notice that the scientific and cultural constructs are falling away from each other; instead, they content themselves with cultural constructions as circulated by the popular media and vested interests.

Even if the overwhelming majority of social and cultural scientists I came across in recent years, continue to close their minds to a transdisciplinary[6] approach to the topic of human-made climate change, however, there are nonetheless outstanding successful examples of the required research collaboration with the social sciences. I refer to the exemplary work of the German media scholar Peter Weingart and of the US-political scientist Roger Pielke, Jr. There are certainly others of similar or even greater caliber — but having found their work relevant for my practical task of communicating climate science with stakeholders, I will describe some of their work in the following paragraphs in some detail.

4.4.1 *The Honest Broker of knowledge*

In his book *The Honest Broker*, Roger Pielke, Jr. (2007) has constructed a provisional typology of scientists. Further, he has described how politics degrades science to a theatre of war by proxy, in order to solve problems that the political system itself cannot solve — no more than science can.

Pielke differentiates five types of scientists, who enter into communication with the public in various ways. The "pure scientist" is essentially driven by curiosity, and has hardly any interest in seeing his new scientific insights placed into a social or political context. The "scientific arbiter" enables an accurate understanding of indisputable scientific facts. Both types are well suited to those situations when a "normal" science can answer questions with great certainty, and these answers, if they come to be socially implemented, are as a rule uncontroversial.

But as I have just explained, at present climate research is not "normal," but rather "postnormal." As a result, we often see the "issue advocate," who applies his scientific competence not to the impartial[7] extension of knowledge, but rather to promoting a value-oriented, and thus also political, agenda. This means that the results of scientific insight are narrowed down to a few, or even to only one, "solution" consistent with his values. The last few decades in particular have produced many scientists of this type, who work and speak for economic or (socio-)political interests. The fourth type of scientist, and the one that Pielke clearly sees as a model to emulate, gives his book its name: "The Honest Broker." This type

[6]In the sense of a collaboration between natural sciences on the one side, and social and cultural scientists on the other side, or more generally a cooperation beyond the modern division of disciplines.

[7]This wording does not imply that strict impartiality would be possible. Certainly, all scientists exhibit or suffer from, some partiality, but most try to constrain this subjectivity to the limited extent possible. Indeed, the building of two opposing separate constructions, scientific and cultural, in this essay represents an oversimplification which helps to work out the argument more easily.

distinguishes himself in that, unlike the "issue advocate," he broadens the scope of the deductions he draws from his findings, rather than constricting it. Thus he enables the political process to choose the "solution" that society desires (and not that which is favored and promoted by the issue advocate). The fifth type is the "stealth issue advocate," who performs the functions of an "issue advocate," but who cultivates the image of an arbiter or honest broker. Due to his fraudulent self-representation, in essence, he benefits neither science nor society.

Pielke recommends that science choose the path of the "honest broker," who explains the complexity of the problems and contributes to weighing up the implications of possible decisions. In doing so, he puts society in a position to choose solutions for its controversies, even on the basis of uncertain knowledge regarding the connections and possibilities, but rationally, and in a manner consistent with its values — for instance, in order to deal with the prospect of climate change that society itself has caused.

The other question is that of the proxy battlefield. Again and again, we see situations in which politicians run aground, coming to decisions that are perceived negatively by significantly large or influential groups. In this case, it happens that a factual constraint is built up, so that policy-makers, in accordance with the scientific analysis, purportedly can come to only one decision. Politics then portrays itself as subordinate to science. This is the case particularly in the field of climate policy, where the "2-degree goal" for avoiding catastrophic climate change, as formulated by scientists, is depicted as an *ultima ratio* to which policy-makers simply must yield. In accordance with Stehr's rule that nothing is as practical as a good theory, because it guides the action to be taken, this depiction is indeed extremely useful politically, precisely because it does indeed guide the action to be taken. Further discussions are not required; the goals of climate policy will be met by means of energy policy, the concept informs. The problem is, however, that the confrontation has been transferred from the visible political stage to the less visible scientific discussion. In that realm, just as little consensus has been reached among scientists as among the politicians, and the resulting argumentative conflict among the scientists degenerates into a political confrontation, fought according to the rules of politics, and ultimately "won" by one party or another.

This process is useful to policy-makers — after all, they come to their decisions more easily — but science is done an injury by thus being politicized. This is not a sustainable use of the resource we call "science," whose service to society by rendering an interpretation of complex facts is ultimately barely distinguished, in the public perception, from the political information disseminated by interest groups.

From this, Pielke derives two normative demands: Namely, that the responsible scientist should act as an "honest broker," and that policy-makers should concentrate on posing only scientifically solvable problems for science, but not on evading their own responsibility to find a "solution" consistent with society's values in normatively difficult situations.

4.4.2 *Risks of communication*

In their book, Peter Weingart and his colleagues (2002a, 2002b) have reconstructed how the topic of climate moved from science to the realm of politics and the media in Germany.

Initially, within scientific circles, there was a phase of "anthropogenization and politicization," according to which human beings were to blame for climate change in the first place, and they could also guide and manage this change by means of responsible behavior. Those responsible, those affected and the options for action were clearly named. Thus, as it was put in the 1986 declaration by the Energy Study Group of the German Physical Society: "in order to avert the threatened climate catastrophe, we must now begin drastically to reduce the emission of trace gases."

This description quickly found its way into the political discussion, because it was also suited to a broader environmental policy discourse. Thus, the concept of catastrophe, once brought into the world, was taken up into the political language. At the same time, the "climate catastrophe" and the struggle to avert it came to be understood and described as an object of policy-based regulation.

The topic was picked up by the mass media, resulting in a further dramatization and intensification. Here, Weingart describes the elements of the "manufacture of climate change as an event," "the staging of the relevance of climate change for day-to-day life," and finally the transformation of the scientific hypothesis into the "certainty of the coming catastrophe."

Peter Weingart substantiates all of these steps by means of examples. Then he poses the question of the risks involved for the three actors: Science, politics and media.

For science, the principal risk is the "loss of credibility due to the particular dynamic of the catastrophe metaphor." This concept enabled climate research to enter onto the stage of politics and the media, but due to its parallel cultural construct, this communication also smuggled along with it a number of connotations. Science is now confronted with these connotations, in statements such as: You made this claim and that claim, how does that match up with this current development or that one? Klaus Hasselmann analyzed this phenomenon in his response, "The Moods of the Media," (Hasselmann, 1997) and lamented that the evidence science was presenting would first undergo a metamorphosis, and science would then have to let itself be measured by these now altered messages. This certainly is not fair, but it is political and social reality. Or, as a journalist once said to me: "Whoever hitches a ride upwards with the media, will meet them again when going down." In both cases, the elevators function according to the same rules.

The risk for policy-makers is in the possibility that the goals set in this manner cannot be achieved. Weingart speaks of a "loss of legitimacy due to taking on too much." That Kyoto was unable to prevent the perceived "climate catastrophe" was foreseeable from the outset; that focussing one-sidedly on energy policy was indeed useful in staging the event for the public, but did not do justice to the facts of the matter.

The media primarily fear the "loss of public attention," due to concepts and conceptual fields becoming worn out. In 2005, when it was declared that there remained only 13 years more in which the climate could be rescued (McCarthy, 2005), and in the following years very little happened, other than rhetoric and symbolic acts on both the scientific and the political side, then the media, for their part, attempted to gain the public's attention by other means; for example, by propagating a skeptical counter-discourse (SPIEGEL, 2007). This is exactly what we observe in the last few years. This counter-discourse, staged by the media, follows the logic of Hasselmann's "Moods of the Media," but also corresponds to the attempt within scientific circles to limit the acceptance of the misleading cultural construct in favor of the more realistic scientific construct. Weingart and his collaborators (2000a) describe this as follows: "The object and trigger of this [climatic] skepticism are, not least, the correcting and relativizing of scientific climate scenarios by established climate research itself. In the sciences this is a normal process; in the media, it becomes an incentive to mistrust."

Science, or more precisely, the scientific institutions, react to this risk by implementing professional "press relations" — which are oriented to "representational principles of the mass media." Policy-makers protect themselves by creating a "hierarchy of knowledge, or of advice," with advisors to the Chancellor, Climate Service Centres and the like. The mass media seek the attention of the public by selectively presenting scientific findings that either agree or conflict with the cultural construct or else by staging controversies, by which means yet another cultural construct is served; namely, the construct of the allegedly arbitrary nature of scientific evidence.

4.5 Concluding remarks

This section is a personal account of a climate scientist, who is often engaged with the need of communicating with stakeholders and with the media, mainly in Germany. During this communication, I was confronted with a number of problems, which obviously are not specific for my own practice, but which I found to be similar with those of other natural scientists.

The key element of my summary is the insight that we have two competing knowledge claims, or, to formulate more precisely, two classes of knowledge claims. For cultural scientists, this is no surprise, but for many natural scientists, this statement is almost an affront.

Scientific knowledge, in social practice, is only one form of knowledge; it must compete with other forms, and it will not automatically "win this competition" — or maybe better: "be accepted as superior knowledge." What would be the "best" outcome of such an encounter? — "best" in the sense of instituting the most rational available understanding of phenomena and their dynamics. It does not imply the availability of immediate practical advice for designing policies, but the presentation of a solid basis for the natural science issues of such policies. A separation of science and values, to the extent to which this is possible. If this goes along a nudging of the cultural construction towards the scientific construction, a gradual rationalization, that would also be a favorable cultural result — for me.

The outlook of this discussion is related to my unfortunate limitation of drawing mostly on Central and Northern European experience. Plausibly, in other cultures, say in Asia or Africa, different attitudes and constructions will prevail and compete with scientific views. This may be in particular so in the US and China. We urgently need cultural scientists and scientific platforms, which bring the natural and cultural approaches together. An example of such a platform is the distinguished journal *Environmental Science & Policy* which was open-minded enough to allow me to publish my ideas.

If scientific actors do not recognize this dynamic, they often attempt to "optimize" the dissemination of their "message" by means of propagandistic tricks — such as emphasizing, or selectively communicating, information to suit its purposes. As a result, first, the public will be disenfranchised; and second, science, as a socially accepted institution, will be damaged. I consider it our task to pursue science on a sustainable basis.

The insight of two competing types of knowledge has a number of practical implications for science. One is, that science itself is under permanent influence of non-scientific knowledge claims, such as ideological or pre-scientific claims.

They influence the scientist in his way of asking and in her request for evidence before accepting answers. Claims, which are consistent with culturally constructed knowledge, are easier accepted as accurate than results, which contradict such claims. Another issue is the transfer of scientific understanding into the policy process. Here, the scientific understanding should help to prepare policy designs — which must not be misunderstood as enforcing certain designs — by clarifying the natural science part of the issues.

Acknowledgments

Paul Malone has done an initial translation of the German manuscript. The editor Jim Briden and an anonymous reviewer have helped considerably with their comment to improve the manuscript. Discussions with Jerry Ravetz, Silvio Funtovicz, Roger Pielke and Peter Weingart were most helpful.

References

Bray, D. and H. von Storch, 1999: Climate science. An empirical example of postnormal science. *Bulletin of the American Meteorological Society* 80: 439–456.

Brikowski, T.H., Y. Lotan, and M.S. Pearle, 2008: Climate-related increase in the prevalence of urolithiasis in the United States. *PNAS* 105: 9841–9846.

Fleck, L., 1980: *Entstehung und Entwicklung einer wissenschaftlichen Tatsache: Einführung in die Lehre vom Denkstil und Denkkollektiv*. Suhrkamp Verlag, Frankfurt am Main, 190 p.

Funtowicz, S.O. and J.R. Ravetz, 1985: Three types of risk assessment: A methodological analysis. In C. Whipple and V.T. Covello (eds): *Risk Analysis in the Private Sector*. Plenum, New York, pp. 217–231.

Hasselmann, K., 1997: *Die Launen der Medien*. ZEIT Nr. 32.

Lovelock, J., 2006: *The Revenge of Gaia — Why Earth Is Fighting Back — and How We Can Still Save Humanity*. Penguin Group, London, 177 p.

McCarthy, M., 2005: Countdown to global catastrophe. *The Independent Online*, 24.1.2005

Milne, A., 1989: *Our Drowning World*. Prism Press, London.

Pielke, R.A., Jr., 2007: *The Honest Broker*. Cambridge University Press.

Pielke, R.A., Jr. 2007: Mistreatment of the economic impacts of extreme events in the Stern Review repon on the economics of c1imate change. *Global Environmental Change* 17: 302–310.

SPIEGEL, 2007: "*Abschied vom Weltuntergang*", 19/2007.

Stehr, N. and H. von Storch, 1999: An anatomy of climate determinism. In: H. Kaupen-Haas (ed.): *Wissenschaftlicher Rassismus — Analysen einer Kontinuität in den Human- und Naturwissenschaften*. Campus-Verlag, Frankfurt.a.M., New York, pp. 137–185.

Tol, R.S.J., 2007: Europe's long-term climate target: A critical evaluation. *Energy Policy* 35: 424–432.

Weingart, P., A. Engels, and P. Pansegrau, 2002a: *Von der Hypothese zur Katastrophe*. Leske + Budrich Verlag.

Weingart, P., A. Engels, and P. Pansegrau, 2000b: Risks of communication: Discourses on climate change in science, politics and the mass media. *Public Understanding of Science* 9: 261–283.

Welch, F., 1995: Me and my God. *Sunday Telegraph*, 10.9.1995.

Yohe, G.W. and R.S.J. Tol, 2008. The stern review and the economies of climate change: An editorial essay. *Climatic Change* 89: 231–240.

Chapter 3

History of Ideas of Climate

The question if and how the climate would influence people and their welfare is an age-old one — at least in the Western culture. The classical answer was climatic determinism (Sections 1 and 2), which has the political advantage that it allowed to consider one's own society as favored by climate, whereas the barbarians in other parts of the world would be disadvantaged by their climate. Proponents of the "scientific" version of this doctrine, such as the geographer Ellsworth Huntington (1876–1947), tried to create an objective fundament of climate determinism in the early part of the 20th century. Stehr and von Storch note that the present-day discussion about the future of humankind as impacted by climate change takes, from time to time, the tradition of the climatic determinism on board.[1]

In the 19th century, geographers began to study the climate in a quantitative manner, based on meteorological observations of the past. A key protagonist was Eduard Brückner (1862–1927)[2] (Section 3), who claimed to have detected quasi-cyclic climate variations, with a period of about 35 years, which would be of natural, probably cosmic origin. This was significant because at that time early claims were made that human conduct, in particular, deforestation could lead to climate

[1] N. Stehr, H. von Storch: "Rückkehr des Klimadeterminismus?" Merkur 51 (1997): 560–562.

[2] N. Stehr, H. von Storch (eds.): *Eduard Brückner — The Sources and Consequences of Climate Change and Climate Variability in Historical Times*. Dordrecht: Kluwer Academic Publisher (2000), ISBN 0-7923-6128-8, 338 p.

change — engendering a public debate with some similarities to the present debate (even if with much less public and scientific attention).[3]

Later von Storch and Stehr discovered that the thesis that humans would change the climate has a much longer history (Section 4).[4]

[3] N. Stehr, H. von Storch, M. Flügel: "The 19th century discussion of climate variability and climate change: Analogies for present day debate?" *World Resources Review* 7 (1996): 589–604. N. Stehr, H. von Storch: „Ein globales Phänomen. Diskussion um Klimaschwankungen vor hundert Jahren," *Frankfurter Allgemeine Zeitung* (1995), 29. März 1995.

[4] H. von Storch, N. Stehr: "Climate change in perspective. Our concerns about global warming have an age-old resonance," *Nature* 405 (2000): 615.

1. Climate Works: An Anatomy of a Disbanded Line of Research*[,5]

Since man is not an independent substance, but rather is connected with all elements of nature; he lives off the breeze of air, as well as of the various inhabitants on earth, food and drink: he uses fire, absorbing light and contaminating the air: awake and asleep, at rest and in motion he contributes to the alteration of the universe and should he not be changed by that same universe?

Johann Gottfried Herder, 1794:87[6]

Abstract

This chapter is designed to advance the view that it is important to move, particularly within social science discourse away from the notion that *climate works* to the idea that *climate matters*. It is in this context that the significant tradition of climatic determinism is crucial. Today climatic determinism lives a strange double existence: It is a widely accepted view among many segments of the public but also among natural scientists; in the latter instance, based on what are taken to be almost self-evident sets of "facts" and in the former case, common sense traditions. On the other hand, among social scientists, it is considered a long-discredited approach that has even less appeal than the notion of inherited intelligence as a basis of social inequality. Whether one is therefore able to move from the compromised notion that climate works to the progressive conception that climate matters is, at the same time, an exercise that forces us to re-open the apparently sealed question of the linkages between natural processes and social action and social conduct and nature.

*This section originally appeared in Stehr, N. and H. von Storch: "Climate works: An anatomy of a disbanded line of research," pp. 137–185 in Heidrun Kaupen-Haas and Christian Saller (eds.), Wissenschaftlicher Rassismus. Analysen einer Kontinuität in den Human- und Naturwissenschaften. Frankfurt am Main: Campus, 1999.

[5]We thank Robert Antonio, Kevin Haggerty, Gerd Schröter, Jay Weinstein and three anonymous reviewers for their constructive criticisms and comments of an earlier draft, which aided greatly in revision.

[6]Authors' translation.

1.1 Introduction

At the present time, the perspective of climate determinism lives a strange dual existence: On the one side, it is, if we are not mistaken, a widely accepted view among lay-people and natural scientists based on common sense traditions and obvious "facts," and on the other hand, among social scientists, it is considered a long and deservedly discredited intellectual perspective.[7] Both views co-exist in a mostly dogmatic and dramatically separated manner. In the decades after World War II, the idea of "climatic determinism" appeared as a sometimes naive, simplified view of the world, and little intellectual energy was put into the topic by serious scientists let alone decision makers in the political arena and the economy. It is only in recent years that the idea enjoys somewhat of a renaissance and mostly in the form of a re-invention of an old idea. From a theoretical point of view, much of modern climate impact research is pure climatic determinism, but the re-inventors of this line of research are mostly unaware of their intellectual predecessors (Stehr and von Storch, 1997).[8]

It is not only in order to avoid the misinterpretations and misconceptions of earlier climatic and environmental reductionism that we suggest that it is mandatory to review the classical climatic determinism concepts; it is also imperative to examine the pitfalls of climate determinism at a time when demands for a

[7] Climate determinism may be a discredited as well as virtually abandoned line of research and perspective in social science, but it is by no means entirely absent from contemporary social science discourse; in support, one could refer, for example, to the climate-based theories of racial identity advocated by Leonard Jeffries (see *New York Times*, Late Edition, East Coast, January 4, 1997, Section 1, Editorial Desk, p. 22): Jeffries "taught a climate-based theory of racial identity in which Africans were cast as 'sun people' who were 'communal, cooperative and collective', while Europeans were scorned as 'ice people' who thrive on brutality and destruction." For an explicit example and proponent of climate determinism from today's natural sciences perspective, see Beck (1993).

[8] The historian Arnold J. Toynbee writing in the early seventies of the last century in an introduction to a biography of Ellsworth Huntington advances a somewhat different notion of the process of intellectual influence, namely as capable to advance and champion ideas in almost unconscious manner or in which amnesia (see also Gouldner, 1980) with respect to work of an individual scholar is not a serious barrier to its cognitive authority: "Huntington is influencing present-day thinkers even if they are not aware of this, and also even if they are aware of it but dissent from Huntington's ideas." Ironically, Huntington himself displays a remarkable degree of his own amnesia when it comes to the great ancestry of climate determinist; for example, in what evidently constitutes the synthesis of his life's scholarly work, in his book *The Mainsprings of Civilization* (1945) one does not find a single reference to Montesquieu, Herder, or Virchow. One can only surmise that he did not see himself standing on the shoulder of giants but rather assumed that his scientific approach obliterated his intellectual predecessors.

revision of the deep intellectual divisions between the *natural and the social science*, brought about last but not least by urgent environmental problems, seem to gain credibility and urgency.[9] It seems to us that an examination of the legacy of climate determinism is even more important in light of the necessity to re-examine the status of "*nature*" in social science discourse.

Our chapter is therefore, first, an attempt at recovering the substance of the classical ideas, their methodological conceptions, and epistemological preoccupations. Second, we are also interested in the lessons the abandoned line of research has to offer for present-day theoretical work and research on the role of natural conditions in human affairs. Such a concern goes beyond the more common and almost uncontested acceptance among present-day social scientists that one of the important but still neglected desiderata of social theory is *the impact of society on the environment*.

In order to advance our agenda, we concentrate mainly on one representative of modern "environmental determinism," namely the geographer and "climate scientist" Ellsworth Huntington, probably the most famous American geographer of the first part of the twentieth century who also was most influential in the scientific community at large and seems to have had a significant sway on the social and political elites in North America as well. Our interest is not so much on Huntington as an individual but on an exemplary representative of a once highly visible intellectual paradigm, now discarded by social scientists.

1.2 The career of a major perspective

For centuries, scientists, intellectuals, humanists, philosophers, physicians, and perhaps the public at large around the globe had few if any serious doubts that climate works. The subject was first discussed, as far as we know, by the physician Hippocrates of Cos (c. 460–470 B.C.), in his treatise on "Airs, Waters, and Places." Although he was primarily concerned with the relation between environment and the pathogenesis of diseases, he digressed into an often-repeated discussion of the effect of climate upon the physical characteristics and the socio-political

[9] The Report of the Gulbenkian Commission on the Restructuring of the Social Sciences Open the Social Sciences (Wallerstein *et al.*, 1996:76) dating from the mid-nineties does not, in our view, pay sufficient or satisfactory attention to this issue although it is mentioned as the question of "how to reinsert time and space as internal variables constitutive of our analyses and not merely unchanging physical realities within which the social universe exists." The problematic of reformulating the location of natural processes in social science discourse goes beyond reflecting on the importance of time and place as does the question of the role of the social in natural science discourse where it now occupies a slum dwelling.

tendencies of the inhabitants of immediate and distant regions. Not much later, Aristotle found a climatic cause for the superiority of the Greeks over the barbarians and therefore for the typical preeminence of one's own climate when compared to that of other places.[10]

For the time being, the career of climate determinism as major intellectual perspective within the social and the natural sciences reached its apex in the first two decades of this century as naturalists, anthropologists, sociologists, physicians, and geographers fashioned a much more quantitative and therefore "scientific" approach to the question of the fateful influence of the natural environment on human civilizations and history. Some of the most definite and assertive statements of climate determinism were published at this time, although in the end they only reiterated convictions held for centuries. Ellen Churchill Semple (1911:1–2), for example, opens her widely cited study on the control of the natural environment over human affairs in 1911 with the following general declaration:

> *Man is a product of the earth's surface ... the earth has mothered him, fed him, set him tasks, directed his thoughts, confronted him with difficulties that have strengthened his body and sharpened his wits, given him his problems of navigation and irrigation, and at the same time whispered hints for their solution. ... Man can no more be scientifically studied apart from the ground he tills, or the lands over which he travels, or the seas over which he trades, than polar bear or desert cactus can be understood apart from its habitat.*

The field of academic geography was a turning point. It was shifting from exploration to explanation. What geography today considers to be discrete forms of discourse, for example, the then closely allied moral and climatic discourses began to be differentiated.[11] At the time, the doctrine of environmental determinism, now often treated as part of geography's distant and disreputable past appeared to offer a solid, broadly-based and scientific foundation serving as the primary explanatory principle of the nature of the interaction between environment and people. The assertion that northern Europeans, in the words of Ellen Semple (1911:620), were "energetic, provident, serious, thoughtful rather than emotional, cautious rather than impulsive," took on an even more pronounced authority that had few if any rivals in the social science community — and the public at large. More recently, contemporary examinations of the relation between

---

[10] A summary of many similar statements over the centuries can be found in Barnes (1921).

[11] David Livingstone (1991) offers an examination of the close linkages among climatic, moral, scientific, and sermonic discourses in 19th-century geography and therefore the common place depiction of the world's regional climates, of race and place cast in moral idioms.

environment and society by social scientists often fail to even mention the rich history of discussions in science that pertain to the influence of environmental conditions on society.

Does that mean that what was perhaps the greatest impetus for the analysis of the influence of climatic conditions on human conduct, namely the desire to arrive at broad even sweeping explanatory frameworks capable of accounting for vast differences in the development of human societies or, using a concept in greater usage decades ago, human evolution, has been widely excised from social science? Indeed, this is the case as far as mainstream social science is concerned. But the principal challenge as seen by geographers, philosophers, anthropologists and many other scientists at the turn of the century was to shed light on what Huntington (1927:136) calls the "degree of progress in different parts of the world."

The interpretative approach of choice that prevailed in response to the intellectual challenges identified by Huntington and others is an "essentialist" perspective; that is, a theoretical platform that assigns inherent, context-independent core properties to climate asserting its definitive influence (power) over all attributes and phenomena of a particular situation. The accepted interpretive convention of climate determinism denies that the "logic" of the situation seen as largely transitory and epiphenomenal can possibly be of any explanatory importance. The essentialist perspective allocates supreme efficacy to climate; and, therefore, "climate works."[12] We propose instead that a host of other factors in our relation to the (natural) climate must be taken into account. One of these conditions, the role of climatic extremes, will be examined.

The fascination, in addition, with the notion of periodicities, cycles, and rhythms of various sorts both as an explanation for the rise and fall of geological phenomena, of plants and animals as well as social and economic processes but also its mere detection was still a vibrant enterprise in science. The conviction that "the whole history of life is a record of cycles" (Huntington, 1945:453) was by no means an idiosyncratic and isolated observation.[13] The regularity and periodicity with which some cycles or waves are seen to reoccur, for example, business cycles, price revolutions or cycles of reproduction are often taken to constitute the

[12] For a critique of what are prevailing, essentialist interpretative conventions of the "nature" of technology and the social role of technology that is embedded in such an interpretation, see Grint and Woolgar, 1997.

[13] On the contrary, the interest in the study of cycles led to the formation of a "Foundation for the Study of Cycles" by Edward R. Dewey in 1941 with a distinguished international board of scientists (cf. Huntington, 1945:458). The Foundation exists to this day and claims to have more than 3000 members (https://cycles.org).

explanation of a phenomenon — last but not least because the phenomenon becomes a predictable process: "It will be a vast boon to mankind when we learn to prophesy the precise dates when cycles of various kinds will reach definite stages" (Huntington, 1945:458).[14]

If, in addition, the assumption is advanced that any definite regularities are the result of underlying physical forces and have their origins there, then the explanatory power is assumed to be even more pronounced. Thus, for Huntington (1945:455) life as we know it is influenced by "at least three kinds of physical conditions, each of which has its own cycles. One of these is the weather, in the ordinary sense of the word. Another is the electromagnetic field of the solar system in general and of the earth in particular. The third is the composition of the atmosphere, with its variations in ozone and perhaps other respects."

The search for periodicities as end-in-itself or as a means for an explanation of the transformation of social phenomena in social science has certainly been abandoned some time ago. One has great difficulty comprehending the excitement which went along with the discovery of cycles and the fascination that such accounts apparently held. Most contemporary social scientists, while not necessarily convinced that history has ended, are sold on the idea that historical processes tend to be rather directionless. In other words, we have arrived at the opposite extreme and we have therefore moved from an intensive search for definite periodicities to the careful elimination of any appearance of historical rhythms and cycles.[15]

It would not be too far-fetched to say that geographers and others who contributed to the modern climate determinism literature, that is, research published during the first half of this century, shared with present-day social scientists an interest in global phenomena. However, the growing contemporary literature on globalization contains at best a passing reference to global environmental problems and then mainly as exemplars of the very existence of global phenomena in the contemporary era.

---

[14]The intellectual interest in cycles, waves, and periodicities also, one ought to note, of course corresponded much more closely to actual experience with the ebb and flow of political regimes, war and peace, hunger and prosperity, the periodical surplus or failure of crops and so on well into the twentieth century. Today, with the advent of knowledge societies, we seem to have entered an age in which the observation of such cycles is much less part of our common-sense experience. Perhaps we have entered the end of the age of cycles. Thus, it is not altogether surprising that the search for periodicities and waves has been discredited.

[15]Exceptions of course are possible, as is, for example, the recent study by the historian David Hackett Fischer (1996) entitled The Great Wave; an examination of price revolutions and the "rhythm of history."

Beginning in the last part of the 19th century and even more pronounced in the first few decades of the 20th-century social scientists indeed abandoned any serious interest in the interaction of natural and social factors. Social science was increasingly based on a specific, restricted concept of space and time.[16] Most theoretical and empirical work assumed that existing political boundaries of the nation state fixed the crucial spatial parameters of sociological, political or economic analysis (cf. Stehr, 1994). Many social scientists also abandoned any interest in comparative analysis and the examination of broad historical trends representing what Norbert Elias (1987) has termed a retreat of sociologists into the present. Social scientists, in addition, increasingly concentrated on accounts of individual rather than general phenomena.

1.3 Acquired and/or inherited

The great deliberation about the relative influence of nurture and environment did not bypass the discussions about the effect of climate on human history and society. During the period under consideration, later contributions are written under the strong influence of Darwinian views while discussions prior to the turn of the century were also influenced by neo-Lamarckian conceptions. Contributions that were indebted to a Lamarckian conception, for example, the views associated with physical anthropology in Germany during the 1880s and 1890s, suggested that as humans are transplanted into a different climate their physiology would actually change and that the organic consequences of acclimatization could then be inherited by subsequent generations.

In the end, it does not really matter whether it is an exclusively Darwinian, Lamarckian or an evolutionary perspective that ambivalently mixes the two approaches[17] and that provides the biological foundation for climate determinism

[16] Some of the disciplines in the humanities, particularly history, did not immediately nor in all countries conform to the restrictive discourse advocated and practiced in sociology and in economics. One of the most notable efforts in history to bring together environmental factors and history may be found in the work of the Annales School in France. The journal *Annales: Economies, Sociétés, Civilisations* that gave the name to the School was founded in 1929 by Lucien Febvre and Marc Bloch and has included among its more recent prominent contributors Emmanuel Le Roy Ladurie and Fernand Braudel. However, there is not a single reference to their work or to that of the geographer Paul Vidal de la Blache who rejected any crude geographical determinism in the early part of this century — and who is considered to be an intellectual forerunner of the Annales School — in the writings of Ellsworth Huntington, as far as we are able to tell.

[17] Herbert Spencer ([1887]:349–350) shares such ambiguity which appears to be typical of contemporary discourse; for example, with respect to the importance of climate he expresses the view that

because these perspectives share the prior commitment to the idea that the natural climate is a basic environmental force that accounts for different manifestations of human success or failure.[18]

The physician and anthropologist Rudolf Virchow ([1885] 1922:231) who espouses a neo-Lamarckian perspective on climate and writing at a time when colonial expansion is on the political agenda, for example, is convinced that the fertility of individuals who migrate to regions of the world in which climates prevail that are different from their "native" climate will experience a dramatic, constant decline in numbers. At least in the short run, the population of colonizers is bound to decrease and can only be sustained as a result of a constant influx of new individuals.[19]

In general, neo-Lamarckians have of course a more "optimistic" outlook in that they are convinced that climate can be conquered almost perfectly by way of adaptation and then inheritance. Darwinians are resigned to the fact that inherited

---

"men having constitutions fitted for one climate, cannot be fitted to an extremely different climate by persistently living in it, because they do not survive, generation after generation. Such changes can be brought about only by slow spreadings of the race through intermediate regions having intermediate climates, to which successive generations are accustomed little by little. And doubtless the like holds mentally. The intellectual and emotional natures required for high civilization, are not to be obtained by thrusting on the completely-uncivilized, the needful activity and restraints in unqualified forms: Gradual decay and death, rather than adaptation, would result." (See also Huntington, 1907:15.)

[18] See the account of the racial myth underlying the settlement of Southern California at the end of the last century in Starr (1985:89–93); Starr describes the conviction of many in contemporary Southern California that it represented the "new Eden of the Saxon homeseeker" and that the Anglo-Saxon stock — weakened by an overlong confinement on the crowded and chilly British Isles — would be reinvigorated and reinforced as a result of the healthy climate in the Southland.

[19] Ellsworth Huntington (1915:6) concurs with Rudolf Virchow and claims, referring to the "poor whites" who have settled on the Bahamas that "when the white man migrates to climates less stimulating than those of his original home, he appears to lose in both physical and mental energy." A more explicit statement that resonates closely with Virchow's observations can be found in a sociology reader to which Huntington (1927b:257) contributed: "If the white man tries to reside permanently on the equatorial coasts of Africa, and to work there as at home, he can scarcely succeed unless his physique is different from that of the average of his race. He must be more leisurely than at home, he must pay more attention to health, his wife and children must often live in more bracing climates if they are to preserve their health. His ideals of public service, of social and scientific progress, and of democratic government may remain unchanged, but lack of surplus energy, even without specific disease, generally causes him to be relatively inactive along such lines. Thus, although the outward forms of society may remain the same in a tropical climate as in more bracing regions, the actual mode of life is almost certain to be decidedly different."

climatic dispositions cannot simply be altered from one generation to another but are at best in a long-term process of natural selection.

Darwinian climate determinists will stress the extent to which climatic conditions attract and pull in some while rejecting others. Similarly, climatic conditions will assert their superiority and drive out cultural practices that are not in accord with them (cf. Huntington, 1945:610). In the long run, as Huntington (1927:165) observes, "ill health, failure and gradual extinction are the lot of those who cannot or will not adapt themselves to the climate, but before that happens many migrate to other climates better adapted to their physiques, temperaments, occupations, habits, institutions and stage of development."

1.4 Ellsworth Huntington

> ***Students of human affairs may agree or disagree with Huntington, but in either case they will be influenced by him, so it is better that they should be aware of him.***
>
> *Arnold J. Toynbee, 1973:ix*

The work of Ellsworth Huntington (1876–1947) on the linkage between climate and societal transformation has to be at the center of this analysis. While other modern-day climate determinists may have changed their opinion or tempered their views on the nature of the impact of climates on social conduct,[20] Huntington really never wavers an inch even when faced with devastating criticism that is supposed to undermine, even destroy his entire argument and line of reasoning (cf. Olmstead, 1912; Sorokin, 1928).[21] As far as we can see, a reference to the work of

[20] Ellen Churchill Semple (1931:99) whose work we cited already and to which we will refer again as a contribution to climate determinism, later developed doubts about her own position and even calls Ellsworth Huntington an incompetent authority on climate who lacks specialist knowledge in climatology; see also Marston Bates' (1952:119) characterization of his own book on tropical climates as an "anti-Huntington tract."

[21] In a letter addressed to Pitirim Sorokin (1989–1968) who at the time was teaching at the University of Minnesota, Ellsworth Huntington (December 15, 1927; Sorokin Papers, University of Saskatchewan Library) refers to Sorokin's forthcoming book on contemporary sociological theories. He indicates that "I am glad that you have thought my theories worth discussing in your new book. I shall not be disturbed by any criticisms which you raise. I am perfectly aware that I often find far more value in books which I severely criticize than in those where the only criticism is that they do not break new ground." His biographer Geoffrey J. Martin (1973:241) reports that Huntington "was annoyed with the type of extended and careless criticism amounting to dismissal which Sorokin offered in Contemporary Sociological Theories." The historian A.T. Olmstead's critique that the phenomena Huntington ascribes to the efficacy climate in fact are the result of cultural factors, predates Sorokin's

Emile Durkheim cannot be found in Huntington. His prolific writings, his considerable fame and wide influence both in the scientific community and in early twentieth American society were remarkable.

Ellsworth Huntington was born September 16, 1876, as the son of a Congregational Church minister in Galesburg, Illinois. He attended high school in Maine and Massachusetts. As a young man, and later in life as well, Huntington traveled extensively in an expedition-like fashion in the Near East, Asia, Europe, Africa, and the Americas. Huntington obtained an undergraduate degree from Beloit College in Wisconsin; he attended Harvard University as a graduate student studying geomorphology or physiography as his teacher William Morris Davis called the study of the form of the earth's surface; in 1907 he joined Yale as an instructor in geography. Yale awarded him a Ph.D. in 1909 and promoted him to assistant professor in 1910. However, in 1915 he was let go only to join Yale again in the autumn of 1919; but he did not teach geography to undergraduates and spent twenty-eight years as a professorial research associate. Our interest is directly in Huntington the human geographer, social scientist, climatologist and historian but not his very public role as a proponent of eugenics. Huntington, the author of *The Goal of Eugenics* (1935) was a significant force in the American eugenics movement and served as president of the American Eugenics Society (1934–1938). There can be little doubt that his work on climate and civilization led to convictions that prompted him to opt for eugenics and become a leader of the eugenics movement in the United States. The bridge is the role of biological inheritance. The euphemism employed to described Huntington's considerable interests in eugenics is that he was concerned "with the quality of people";[22] according to Huntington "democracy itself was threatened by the rapid multiplication of the less able members of the species" and he urged "restrictive immigration into the United States" (Martin, 1973:xiv).

As a matter of fact, Huntington was generally driven by a concern about concrete ways of improving human existence and he rarely hesitated to make practical suggestions and offer policy advice in the area of climate matters as well.[23]

---

indictment of Huntington later work by decades. Olmstead (1912:168) concludes his critique by saying that "the historian is not justified in utilizing climate for more than the study of the background of his history. For influence on particular events, there are many geographical facts of far more significance." Huntington's response to Olmstead may be found in his essay "Changes of climate and history" published in 1913 in the *American Historical Review*.

[22] A term Huntington (e.g., 1945:313) himself employs.

[23] In an effort to limit our examination of climate determinism, we decided not to investigate the varied political, ideological and industrial uses made of climate determinism in different contexts. Perhaps in accordance with the claim that environmental determinism is an eminently scientific

Huntington is rarely satisfied with merely documenting his case, he also desires to draw practical lessons and apply his conclusions at once. The practical clues came directly from his research. Huntington wants us to exploit the benefits of climate: For example, he suggested that the seat of the United Nations should be located in Newport, Rhode Island because it had the most suitable climate for humans. And his concern for the optimum (indoor) climate even resulted in a close association with the American Society of Heating and Plumbing Engineers (cf. Martin, 1973:xiv).

Aside from publications in geography that are essential textbooks and his writings on eugenics, Huntington's main work relevant here concerns the reasons for the "progress" of human civilizations. These ideas took shape first around 1914–1915 and quickly were crafted into a definitive thesis that changed little during the subsequent three decades. And as the historian David Arnold (1996:31) has recently indicated, "like so many environmentalists before him, he first looked east, seeking in climate and climatic change an explanation for the differences between Western dynamism and Eastern stagnation."

Huntington's early writings on climate change in post-glacial times surely were stimulated by an interest among the fields of geography, climatology and geology that existed in Russia, Germany, Austria and other countries in the phenomenon of climate changes in historical times: This peaked at the turn of the century and quickly gave way to the conviction that climate is essentially a static phenomenon. Whether the long-term changes that were identified merely reflected recurring oscillations, fluctuations or "pulsations," or were indicative of progressive changes toward different states of the climate, for example, aridity, were among the contentious issues in discussions among scholars who shared the conviction that significant transformations could be observed. The subsequent and remarkable change in emphasis in disciplinary paradigm or tradition is also reflected in the work of Huntington. In his early work, there is a distinct emphasis

---

undertaking and, therefore, must have an undeniable powerful practical utility, proponents of environmental determinism never hesitated to underline its eminent usefulness. Huntington, for example, between 1921 and 1929 was chair of a National Research Council (NRC) *Committee on the Atmosphere and Man* (CAM) that from its inception focuses its work on four projects: "an investigation of the influence of meteorological conditions on factory productivity, physiological experiments under laboratory conditions, experiments in hospitals, and an investigation of mortality caused by influenza in New York City" (Fleming, 1999). As a result, and with good justification, Fleming calls Huntington's activities in this field as a form of "meteorological Taylorism." In addition, the enlightening case study by Frenkel (1992) of the role of environmental determinism in the development of the Panama Canal Zone very well illustrates the practical efficacy of climate determinism as an intellectual or ideological weapon (see also Weinstein and Stehr, 1999).

on climatic change, variability, and fluctuations. In his subsequent work on climate and weather, and especially in his synthesizing books published in the 1940s, the time-horizon changes and the emphasis moves away from climatic variability and fluctuations and is replaced by accenting the essentially fixed nature of climate. In Huntington's case, the shift of interest away from a concern with longer-term climatic changes and its periodicities also meant that he focused on the cycles of what are essentially weather rather than climate patterns; for example, around 1914 and 1915 Huntington began to assemble and collect a great variety of empirical information about the impact of changes in weather patterns on daily "nervous activity," productivity, "feelings and energy." And it is during these two years and on the basis of such data that Huntington formulated the essence of his thesis on the linkage between climate and human activities. The initial findings were published in a series of three articles in 1914 commissioned by *Harper's Magazine*. The essays under the title of "Work and weather," "Climate and civilization" and "Is civilization determined by climate?" "brought letters of anger from southern gentlemen, inquiries from medical men and psychologists, [and] a proposal for civic celebration in Seattle" (Martin, 1973:114). Ironically, with the shift in his interests to the impact of climate on human affairs comes the retreat from the analysis of climate change in historical times. The latter could in fact have undermined the conclusions of his work on the progress of civilizations. Nature now exists in a state of permanence and harmony although the consequences of nature for society are not always beneficial, for that depends on where one happens to live.

1.5 Health, energy and progress

> *No nation has risen to the highest grade of civilization except in regions where the climatic stimulus is great. This statement sums up our entire hypothesis.*
>
> *Ellsworth Huntington [1915] 1924:(first ed. 270)*

Our discussion of *modern* climate determinism is primarily limited to contributions dealing with the profound effects of climate on human life published during the last two decades of the last century and the first few decades of the 20th century, at a time when climate determinism aspired to become a recognized scientific endeavor. Climate for Ellsworth Huntington (e.g., 1945:307) mainly meant attention to temperature and, secondarily, seasons,[24] storms and precipitation. Toward

[24] Huntington (1945:313) formulates with respect to seasons the following generalization: "The greater the contrast of seasons, the greater in general are the demands on man's strength and skill in order to get a decent living."

the end of his scientific career, in his last major book, Huntington (1945:313) sums up his main thesis about the salience and efficacy of climate in the following manner: Climatic conditions constitute a distinct optimum (and conversely, a downside) and with it varies "the advance of civilization and the quality of the people."

According to his biographer, the idea that climate stimulates or inhibits human energy, health and progress appears to have been suggested to Huntington initially by Charles J. Kullmer, a professor of German at Syracuse University who wrote to him in September of 1911 indicating that he had been working on the concept of climate change for some time and sent him a manuscript (cf. Martin, 1973:102). Kullmer indicates in his letter that he had followed Huntington's work, for example, the book on his travels in Asia and that he wanted to consult him on the hypothesis of climatic change and its connection to civilization.[25] In particular, Kullmer claimed to have found a close correlation between storm tracks and civilization and that shifts in storm tracks accounted for shifts in the location of civilizations.[26] What we are therefore confronted with is that many of the basic qualities of life in different regions of the world with similar climates tend to converge while in other respects which "have little relation to physical environment they may differ radically" (Huntington, 1945:611).

On their own, these and related hypotheses appear to support a fairly innocuous thesis acknowledging perhaps nothing more and nothing less than the distinct possibly that natural conditions impinge in various and therefore not fixed ways on human conduct. However, one needs to recall that Huntington is also convinced and attempts to offer massive supporting evidence that the evolution of civilization itself as well as the "quality of the people" cannot be separated and understood aside from climatic conditions that either favor or deter their development. For Huntington, the nature of the civilizational progress either facilitated or inhibited by climate refers to "[our] increasing ability to dominate the forces of nature. ... Is it mere coincidence [he asks] that the English can fly in the air, sail beneath the ocean, manufacture machines by the million, and talk by radio, while not a man

[25] In the introductory chapter to *Climate and Civilization*, Huntington ([1915] 1924:7) relates that he first developed his theories of climatic "pulsations" during the Pumpelly expedition in 1903. In the two years he spent in Turkestan, he came to be convinced that "Reclau, Kropotkin, and others are correct in believing that two or three thousand years ago the climate of Central Asia was moister than now."

[26] Cf. Huntington's (1927:143–145) summary discussion of the importance of storms, as the "third great element in producing changes in weather," in addition to temperature, humidity, and seasons that prove to be particularly valuable for health and climatic energy.

among the Kamchadales ever thinks of doing these things?" (Huntington, 1927:136–137). Interestingly, Huntington's definition of progress as emancipation from the forces of nature does not mean liberation from climate in the course of civilizational progress. On the contrary, progress implies greater and greater dependence on climatic conditions because the "centers of civilization keep moving into the regions where man's stage of progress makes him most efficient" (Huntington, 1927:161). As a result, the direct effect of climate assumes even greater importance.

Today, Huntington's (1927:138) tireless work affirming the thesis that "climate paints the fundamental colors on the human canvas" appears to be amusing to some, to others extreme or lazy (Le Roy Ladurie, [1967] 1988:24), but most would likely consider it absurd and therefore certainly on the very margins of social science discourse about the impact of environmental factors on the human condition. However, in its own time, it was by no means atypical nor did it necessarily contradict commonsense assumptions about climate, health and ethnic or racial identities. Huntington's views were easily assimilated to and resonated with the doctrines of racism and imperialism of his day. Their very success and political utility prove to be a firm condition for its obliteration today. The remaining value Huntington's program may have appears to be limited to its productivity as counterexample. Its particulars, as enumerated in our appendix on the efficacy of climate according to Huntington, perhaps assist in avoiding to be drawn again into reflections on climate that resemble and resonate with the tradition of modern climate determinism.

1.5.1 *Climatic optima and downsides*

Aside from the rich description of psychological, social, economic and political features of human life and society that are seen to evolve — or are hindered from emerging — in response to climatic conditions, climate determinists also advance a list of major climatic factors associated with fortuitous benefits or extreme downsides of climate. There is, first of all, the dreadful *climatic monotony* or, for that matter, perhaps equally feared, *climatic extremes*. In contrast, there is the very beneficial pulsating *climatic variety*: The specific examples for each of these conditions typically are linked to just a single meteorological element, be it temperature, rain, wind, humidity, seasons, etc. The main assertion Huntington (1927:142) advances when it comes to the identification of climatic optima is that "*change*" as such is exhilarating. Huntington's (1927:141–142) enumeration of the conditions that constitute the best climate for human health, progress and energy therefore is somewhat richer since he lists a number of climatic conditions that should be present simultaneously:

1. A fairly strong but not extreme contrast between summer and winter is needed, the summer temperature averaging not much higher than 65°F for night and day together. This appears to be the temperature at which the white race is physically most active and healthy. The winter temperature out of doors should average not much below 40°F, for this is the temperature at which people with our type of food, clothing, shelter, and occupations appear to be most active mentally.
2. There must be rain at all seasons. This does not mean constant rain, but enough so that the air is moderately moist much of the time. If the air is dry for any long period, people's health is not so good as when it is damper. Abundant statistics in many regions demonstrate this in spite of the popular opinion to the contrary. That opinion probably has arisen because people confuse the beneficial effect of the outdoor life in dry climates with the effect of the dryness itself, or of the dust which comes with the dryness.
3. Constant but not undue variability of weather is almost as important as the right conditions of temperature and humidity. Among factory workers and students, for instance, it has been found that if the temperature of one day is the same as that of the preceding day — which generally means that the other conditions are likewise uniform — people's work is not so good as if there is a change, especially a drop of temperature. The point of the matter is that the *change* is exhilarating.

1.5.2 *The foundations*

> *Das Voralpenklima (in Salzburg) macht gemütskranke Menschen, die schon sehr früh dem Stumpfsinn anheim fallen und die mit der Zeit bösartig werden.*
>
> *Thomas Bernhard, [1983] 1988:19*

Once an enumeration of the seemingly endless list of factors and processes that are ascribed to be determined or effected by climate has been made (see our *appendix* summarizing these notions alphabetically), important questions arise about the foundations, be they theoretical, empirical or both that are explicitly advanced to make the case for the importance of climate in human affairs. Early in his discussion of the linkage between climate and social conditions, Huntington invokes and appeals to experiences that he expects surely every one of his readers' shares and can replicate almost perfectly and instantly. Although it is quite easy, as a fellow geographer (Spate, 1952:413–414) remarks in a more recent review of Huntington's work, to point to the anomalies in the famous civilization — and — climate maps, "in our hearts most of us Westerners probably believe that the facts are as stated." That is to say, in Huntington's discourse about climate and civilization we encounter an invocation of traditional cultural and political beliefs. By referring to what Huntington considers to be self-evident as well as a widely shared elementary everyday experience and reaction to changing weather conditions, for example, he means to spawn among his readers a kind of essential assent to his thesis. We want to refer to convictions that every individual shares with every other individual. Thus, the most fundamental evidence Huntington adduces is this appeal to basic and widely shared traditional beliefs or prejudices about otherness and the way in which individuals in different climates respond to climate. Elements of such shared, intuitive evidence are of course linked. After all many climates are variable enough to allow for personal encounters with a range of weather extremes.

Huntington (1920:249) summons everyday experiences such as these: "The variations in people's strength from month to month are so important and teach us so much about the distribution of health and energy throughout the world that we will study them closely." More specifically, therefore, "let us consider how physical strength varies during the course of the year in the great section extending from southern New England and New York westward to the Rocky Mountains. October is usually the best month. At that time people feel like working hard; they get up in the morning full of energy, and go at their work quickly and without hesitation; they walk briskly to business or work; and play with equal vigor. Headaches, colds, indigestion, and other minor illnesses are fewer than at other seasons; there are also fewer serious illnesses, so that doctors have less than usual to do, and the number of deaths is less than at any other time of year." Perhaps with the exception of the very last assertions, these are all observations that appeal to everyday experiences and are assumed to be easily replicated. And the same is true for the conclusion that there is the "well-known contrast between the energetic people [Huntington just described] of the temperate zone and the lazy inhabitants of the tropics" (Huntington, 1920:248). It is inescapable and widely taken for granted that "everyone is influenced by temperature, humidity, wind, sunshine, barometric pressure, and perhaps other factors such as atmospheric electricity and the amount of ozone in the air. On days when all these factors are favorable, people feel strong and hopeful; their bodies are capable of unusual exertion, and their minds are alert and accurate. If all the factors are unfavorable, people feel inefficient and dull; their physical weaknesses are exaggerated; it is hard to concentrate the mind; the day's work drags slowly; and people go to bed at night with a tired feeling of not having accomplished much. Hence in a variable climate like that of the United States, people's physical and mental energy keep changing from day to day and season to season. Sometimes one feels almost as inert as if he lived within the tropics, but soon a change comes and one again feels the health and energy which makes it possible to work hard and think clearly" (Huntington, 1920:248).

At the center of the foundations of Huntington's observations about the work climate clearly is an appeal to what he believes are almost universal and powerful commonsense experiences with weather conditions. He asks us to rely for confirmation of his basic assertion on self-analysis, on how we respond to varying weather patterns or climatic conditions. Huntington is convinced that all of us easily identify with his conclusions because we can quickly and surely assemble experiences that warrant the basic thesis as factual.

1.5.3 *The limits of imagination*

In the end, the type, range and possible limits of kinds of human conduct that are attributed to climate are, so it seems, only limited by the limits of the imagination of the authors. Any superficial examination of the inventory reproduced in our Appendix of forms of social conduct brought about by climatic conditions must conclude that this is an almost exhaustive list of consequences. But this is not really the case. There are discernible limits. And the boundaries are those of the particular theoretical and cultural commitments of the author. Nonetheless, the mere enumeration of factors and processes that are seen as varying with climatic conditions and regions indicates that there are few striking limits. The same conclusion can be drawn from what is an essential lack of discipline or constraint when it comes to what climate determinists claim to be linked to climate. In a dispute about the importance of the role of different explanatory factors, Huntington (1914b:19) feels prompted to make a similar observation, except that it is among the *métier* of the historians that he discovers what might be called the cognitive fallacy of failing to restrain one's assertions:

> *At the beginning of their volumes, the historians speak respectfully of the influence of geographical factors, but that is usually all. Thenceforth they become so impressed with the importance of economic considerations, or of purely human matters, such as ambition, religious ardor, mechanical invention, constructive*

> *statesmanship, or scientific, literary, and artistic achievement, that they feel that other subjects are scarcely worth considering.*

But what about the role of the historians matters if "in many ways" they are "molded by the physical environment" (Huntington, 1914b:19)? At the outset of one's reflection, purely human matters deserve brief mention but as Huntington himself demonstrates, the geographer promptly follows the lead of the historians he criticizes and cultural, merely human matters that is, promptly withdraw into a black box:

> *Among primitive men the nature of the province which a tribe happens to inhabit determines its mode of life, industries, and habits; and these in turn give rise to various moral and mental traits, both good and bad. Thus definite characteristics are acquired, and are passed on by inheritance or training to future generations (Huntington, 1907:15).*

Moreover, it is quite common to find that individual authors will do their best to remain internally consistent, for example, by arguing that Northern latitudes typically go hand-in-hand with such and such temperament and traits. Yet different authors who agree on the extraordinary influence of climate on human affairs, but who obviously do not confer with each other about specific attributes and geographic boundaries within which they are supposed to occur, will often advance entirely contradictory substantive assertions.

While Huntington, for example, insists on the fate-like effect of climate-induced differences between Northerners and Southerners in most countries, Leroy-Beaulieu (1893:139–144), on the other hand, is convinced that there are discernible convergences in the character of Northern and Southern Europeans because the populations in both regions are subject to climatic extremes and long periods of enforced idleness, as a result. The upshot of course is that climate determinism as a whole has as one of its profound characteristics a kind of arbitrariness. Such arbitrariness, of course, dissolves at the level of the individual author. The speculations about the force of climate become an ill-disguised substitute for ideological and ethnocentric beliefs: As widely proclaimed, "temperate climates or 'mild' climates were favorable to the development and survival of a superior type of people, but each writer has construed the doctrine so that his own land was regarded as the norm of the temperateness in climate" (House, 1929:17).

Perhaps other discernible limits and conditions for possible forms of social conduct that are rarely enumerated as "caused" by climate are important as well. We are thinking especially about the absence of any mention of "technology" and technical developments in the climate determinism literature. Huntington does refer as cited to innovation in the field of technology as linked to climate but is silent about the ease of its dissemination. Thus, if modern technology is unprecedented and one of the attributes that separates the last two centuries from all previous history,[27] then the omission of any reference to the global impact of technological regimes may well be significant. For if the uniqueness of today's experience is the uniqueness of the technical and scientific knowledge that gives rise to what is not only the motor of the modern economy but of modern warfare and the conditions for peace, then such a blank is quite significant.

1.5.4 *The power of generalizations*

Among the central features of texts written by climate determinists are not only their almost poetic excesses but also their mundane redundancies. In addition, one of the distinct narrative features of the discourse of climate determinists about the pervasive authority of climate over human affairs concerns the almost runaway vigor of their assertion about climate, for it quickly becomes a powerful and all-exclusive generalization that drives out any qualifications.[28] And to that extent, the narrative again and again literally is immune from efforts to restrain it by alluding to other or "intervening" forces, restrictions or exceptions.

---

[27] Not irrelevant in this context is Werner Sombart's (e.g., 1931:98) description of modern technology as the liberation of the economy, for example, from the limits and constraints imposed by living nature. Climate in this sense is living nature and the emancipation from climate is the liberation from living nature.

[28] See Max Weber's ([1909] 1922) critical review of Wilhelm Ostwald's cultural theory based exclusively on the "energy" metaphor resulting in an energy dependent and energy driven perspective of the development of cultures. Weber notes with dismay that Ostwald is unable to restrain his generalization despite his own best intentions (e.g., Weber, [1909] 1922:387). C. Wright Mills (1959) has written a widely known and often quoted indictment of "grand theory" in sociology. However, in contrast to Grand Theory — as a form of an all-inclusive, hermetic perspective and a style of work — castigated by Mills and described by him as a not readily comprehensible social science discourse, Huntington's generalizations are immediately intelligible and applicable. They do not seem to outrun, as is the case for grand theory, any specific and empirical problem or constitute a formalist as well as permanent withdrawal into systematic work . "The basic cause of grand theory," as defined by Mills (1959:33) "is the initial choice of a level of thinking so general that its practitioners cannot logically get down to observation." Huntington's generalizations need not be "translated" into plain or straightforward English. Mills (1950:31) tries to demonstrate that the result of the translation effort of grand theory would not be very impressive. Huntington's generalizations clearly do not raise the issue of intelligibility. Perhaps they raise the opposite dilemma. They are too impressive. They do not have a sense of "unreality" surrounding them.

Take, for example, Huntington's (1945:275) efforts to confine and delimit his own rhetoric about the utmost significance of temperature on human affairs. He summarizes and concludes the relevant discussion as follows:

> *thus, if all other influences were eliminated, we should expect civilization to advance most rapidly in climates which have few or no months with temperatures above the optimum and many below, but none too far below, the optimum. As a matter of fact, the actual distribution of civilization approaches this pattern but departs from it in some respect because mean temperature is only one of the climatic factors of environment, and the effects of physical environment are modified by cultural environment.*

As far as we know, there are of course no human civilizations anywhere that would enable us to observe their comparative development solely on the basis of non-climatic factors. But this does not really matter since the development of civilizations we in fact are able to observe, corresponds to such close degree to the expected evolution in response to different climates that one can discard, or even ignore, other environmental factors and culture.

Moreover, since temperature is but one among a range of climatic attributes, the correlation between environment and civilizational development is actually underreported as long as one relies for its empirical representation on data about temperatures alone. This in turn considerably strengthens the case that climate is the crucial dimension. In other words, efforts to restrain the generalization about climate often appear to have the opposite effect, they appear to reinforce and invigorate the generalization.

Similarly, as Huntington (1945:344) attempts to explicate the influence of climate on mental activities, in particular in light of what some have called the rise and fall of entire civilizations or the absence of remarkable intellectual accomplishments in regions in which the climate is almost at an optimum, he builds bridges, advocates caution, hints at exceptions, appears to minimize the influence of climatic conditions but in the same context also injects entirely new hypotheses which almost totally eliminate any ability to "falsify" his generalizations. The assertion that we are from time to time faced with major climatic cycles in history is a prime example for such a hypothesis that immunizes assertions almost completely against any falsification attempt. In the end, it would appear that we face an insurmountable argument about the influence of climate on human conduct, an argument in the form of a tautology. Huntington (1945:344) indicates, for example, that mental alertness or intellectual activity — rather ambivalent terms, to say the least — depend on a variety of factors:

> *Climate and weather are simply others in this series. They receive special treatment here because they are little understood as yet and because their cyclic variation seems to have influenced some of the greatest historical changes. The highest mental achievement is possible only when favorable conditions exert a combined stimulus. Our task just now is to separate climatic effects from those of heredity, culture, and the non-climatic physical environment.*

In short, Huntington never lives up to his promise to factor out different influences but construes chains and causal connections among factors, so that, in the end, only climate emerges as the real and effectively independent variable in the equation.[29]

Perhaps the power of the generalization is even more intense because Huntington tends to reverse possible qualifications of the impact of climate on society by suggesting that social forces in the end actually reinforce "climatic destinies." For example, he refers to selective migration that amounts to a kind of climatic cleansing, "a process of selection through migration is tending, slowly perhaps, to concentrate the more easy-going type in the warmer climates" (Huntington, 1945:277). All of this only reiterates again and again the basic insight that "social and economic systems everywhere tend to adjust themselves to geographical environment and to the occupations which provide a living in a particular environment at any particular stage of human progress" (Huntington, 1945:280). The generalization quickly and surely has been stripped of all restrictions and qualifications.[30]

---

[29] One fascinating statistic that Huntington (1945:345) utilizes in this context are circulation figures from twenty-eight public (city) libraries in the United States and Canada (more precisely, weighted averages — generally for twenty years, 1920–1939 — of fiction and non-fiction books in circulation). Huntington groups the libraries into four categories according to latitude. In the six most northerly libraries (St. John, New Brunswick, Minneapolis, Portland, Oregon, Seattle, Spokane and Vancouver, British Columbia) the proportion of all circulating books that are classified as non-fiction is 55.2 while the corresponding figure for the eight most southerly cities (Tampa, Houston, New Orleans, Jacksonville, El Paso, Savannah, Shreveport, and San Diego) is 28.9 percent; actually, the reported differences are confined to the latter group and all other city libraries because the two intermediate categories also have a percentage of non-fiction circulation above fifty percent. Huntington simply reports these figures, of course quite uncritically, and without further comments. He is convinced that the reader will be convinced as well that the evidence is clear-cut and of undeniable force. What the evidence confirms according to Huntington is that people of high latitudes are, on the whole, more intellectual than those of low latitudes.

[30] See also Huntington's (1945:24) discussion of the various maps of the United States he adduces to buttress his argument about the essential superiority of climatic factors as an explanation for a host of features of social life (most of those already enumerated). He pores over these maps and discovers a variety of "minor differences" or a lack of full resemblances with the basic pattern of climatic

The general question these peculiar features of Huntington's discourse raises concerns the reason for his inability to restrain his generalizations despite what we assume are his good intentions to do so. That is, if we assume for a moment that his announced efforts to suppress excessive generalization are well intentioned, not merely a preemptive strike against critics precisely bemoaning the lack of restraint or the result of advocating, in the end, factors that resonate more closely with his disciplinary identification, we have to ask what might account for the difficulties in restraining one's generalizations. After all, this is not a dilemma peculiar to Huntington.[31]

1.6 Why does climate not work?

> *In contrast to the helpless dependence upon environment of stationary plants and animals, whose range of movement is strictly determined by conditions of food and temperature, the great mobility of man, combined with his inventiveness, enables him to flee or seek almost any climatic condition, and to emancipate himself from the full tyranny of climatic control by substituting an indirect economic effect for a direct physical effect.*
>
> *Ellen Semple, 1911:608*

It is peculiar that climate determinists also offer arguments that negate their own perspective. Take, for example, Ellen Churchill Semple's (1863–1932), the first

---

efficiency and then concludes that all the maps really show the same basic feature and that "the resemblances are too close and too widespread to be accidental." The maps acquire their basic resemblance from climate. He adds, "nothing that man can yet do has any appreciable effect upon the weather, with its changes from day to day and season to season, or upon climate, with its temperature, humidity, and wind. On the other hand, everyone knows that human feelings, health, and activity are extremely sensitive to weather and climate" (Huntington, 1945:249).

[31] In a 1940 essay reviewing the work of Earnest Albert Hooton and his imperial assertion about the organic basis of crime, Robert K. Merton and M.F. Ashley-Montagu note the presence of two distinct but concurrent interpretive tendencies in the discourse of biological determinism. Resonating with the similar discursive inclinations in the work of Huntington, Merton and Ashley-Montagu (1940) observe that there is a "cautious and admirably restrained effort to assay the significance of biological factors in the determination of the incidence of criminal behavior; the other, a pugnacious and flamboyant insistence on the biological determination of crime." The cautious qualifications and laudable protestations that crude determinism ought to be avoided are almost always forgotten in the fervor of formulating unambivalent theoretical conclusion as well as decisive practical-political demands, in the case of Hooton, about the utmost significance of organic factors in producing, explaining and responding to criminal conduct.

female president of the Association of American Geographers observations about what Rudolf Virchow (1821–1902) calls the "cosmopolitanism of the human being" (Virchow, [1885] 1922:216), namely the ability of humans to settle in any parts of the world; such an assertion about the "openness" to environmental conditions of humans obviously severely restricts or limits the potential work climate can do.[32]

Without question, climate determinism lacks analytical elegance; it often conflates the "climate variable" with other explanatory factors and borders on the tautological. Some of these features it shares with other grand theories designed to explain civilizational transformations, but what should concern us most is the poor example climate determinism offers for work that proposes to bridge the divide between the cultures of the social and the natural sciences and the potential dangers or misunderstandings "scientific" climate determinism may generate as it enters the public arena.

But in order to indicate why climate does not work in the way in which climate determinists are convinced it does, it is necessary to explicate additional assumptions that typically accompany discourse of climate-based theories. A critical analysis of the assumptions will lead to the conclusion that climate matters but does not work — at least not in the undifferentiated and indiscriminate fashion found in the literature committed to climate determinism.

The assumptions or the climate construct to which we want to draw attention concern the following attributes of discourse of climate-based theories of social conduct: (1) The essential stability of climate and conduct; (2) Climate does not tend to discriminate; and (3) the one-dimensionality of climate. Aside from the features we already have identified, especially the inability to constrain the basic assertion and that climate, as a result, effects human conduct without exception, the assumptions to be explicated now have the remarkable common attribute that

[32] Rudolf Virchow's assertion is in line with and in the spirit of his staunch advocacy of monogenism. In contrast to the United States and France where influential "polygenist" supporters were influential during the latter part of the 19th century, German anthropology during the same period was dominated by scientists who strongly believed in the "unity and equality of mankind." Virchow's radical liberalism generally was well known and included his vigorous opposition to Teutonic racism among anthropologists and in politics; Virchow's position is by no means consistent. Most physical anthropologists before 1900 were neo-Lamarckians. But depending on the circumstances Virchow takes a neo-Lamarckian or Darwinian position. In the case of his discussion of "acclimatization" in 1885, Virchow advocates a staunchly neo-Lamarckian position in opposition to August Weismann's theory of heredity; Weismann had claimed during the same conference that humans are not capable of acquiring and then biologically passing on characteristics in response to climatic conditions (see also Massin, 1996).

they all contradict some of the most widely shared convictions among social scientists about the "nature" of social life. That is, (1) social life tends to be fragile; it is constantly changing and attention to its mutable character is a prime requirement in examining any social action whatsoever, (2) most things in life tend to be stratified, and (3) social conditions tend to be "complex" quite independently of their volume, range and significance. But first, we want to explicate the climate construct employed by Huntington.

1.6.1 *The social construct of climate*

The social construct of climate found in Huntington's writings can be best be described as a meteorological construct. Its scale is regional. The impact of climate is unconditional. As a matter of fact, the operative climate construct is virtually taken for granted and largely obscured in Huntington's writings. For example, Huntington ([1915] 1924:136) approvingly refers to Mark Twain: "Climate lasts all the time and weather only a few days." But what exactly lasts or varies all the time, Huntington does not say. However, based on how he examines the effects of climate on society, it becomes evident that Huntington's conception of climate resonates strongly with and affirms what the pioneers at work at the turn of the century of the emerging scientific fields of meteorology and climatology considered to be climate.

One of the most important meteorologists of the time and considered to be one of the founders of modern meteorology as the science of the physics of the atmosphere is Julius Hann.[33] In the classical *Handbuch der Klimatologie*, first issued in 1883, Hann (1883:1) defines climate as the "sum of all meteorological phenomena that characterize the average conditions of the atmosphere at any given location of the earth." From an operational point of view, and given the technical means

---

[33] Julius Hann (1839–1921), who was born in Wartberg, Austria, studied mathematics, physics, geology and geography at the University of Vienna. After a career in teaching, he became professor of physics at the University of Vienna and in 1897 professor of meteorology at the University of Graz. Between 1900 and 1910, he occupied the newly created chair for cosmological physics at the University of Vienna and served as director of the Institute for Meteorology and Geodynamics. Hann was an enemy of speculative thinking; his main goal was to establish the facts (Brückner, 1923:155). Hann was descriptively oriented, that is, keen to establish the observational basis for various meteorological phenomena. In addition, Hann was for more than fifty years editor of the *Meteorologische Zeitschrift*. He died in 1921 at the age of 83 in Vienna. Julius Hann compiled the first textbook on climatology. He first published his *Handbuch der Klimatologie* in 1883; the *Handbuch* appeared in a number of subsequent editions and became a classic in climatology. An English edition based on the second edition of the German version of the Handbook was published in 1903 (Hann, 1903).

available at the time, Hann's definition of climate refers to macro-meteorological phenomena that can be *measured* at the *surface* of the earth. Climate is the sum total of quantifiable climatic elements especially temperature, humidity, precipitation and wind speed averaged over a certain period of time. As Hann stresses, in contrast to a mere and indeterminate subjective impression of climate, scientific apprehension of climate requires the numerical expression of climate elements based on empirical information. When it comes to ordering the relative importance of the range of meteorological phenomena Hann (1883:5) advocates that their influence on "organic life" should be decisive. Climatology itself is not able to advance such a ranking. It is dependent, for example, on geography. Huntington follows Hann's reasoning to the letter. For Huntington climate without exception means an average of one of the meteorological conditions. In most instances, this happens to be temperature. Hann's discussion of individual meteorological phenomena in his *Handbook* indeed begins with what he considers to be the most important condition, namely temperature. From the beginning, instrumental measurements of the meteorological conditions imply that they constitute and are relevant as macro-phenomena. They refer to conditions that exist outside of what might be called the indoor-climate fabricated for millennia by society for its members, for example, in the form of clothing, nutrition and shelter.

However, based on the widely shared self-understanding and ambitions of climatologists and geographers, the climate conception employed by Huntington could also be called a naturalistic, or scientific conception of climate. It is nature itself, in this case, the dynamics of the atmosphere, that speaks to the observers through the readings of the instruments. There is no indication that we are confronted with a reading of nature that is culturally conditioned. Huntington's confidence in the tremendous impact of climatic conditions on individual, society and civilizations is obviously reinforced by his macro-meteorological conception of climate, for it appears to relentlessly impose its force on humans in an unmediated fashion from which there is no escape. Natural conditions, for example, available natural resources and their limits but also climatic processes do affect human conduct and be it only as the result of certain social re-constructions of these features as constrains of social conduct; but they only constitute constraints for human conduct, they do not necessarily determine it. Even as conditions their impact varies historically, is stratified, at times virtually perceived to be negligent, at times seen as crucial. The same applies to climate. Climate conditions human conduct only insofar as it is perceived and socially constructed as such a condition. It does not affect social conduct in its pristine, objective condition (see also Hoheisel, 1993:137). Climate does not affect us both in its material and cognitive sense unconditionally, as Huntington still believed.

1.6.2 *Climate does not discriminate*

Among the characteristic "social scientific" features of discourse that champions climate determinism is, as one might call it, its peculiar egalitarianism. Climate is responsible, as we have seen, for a wide range of human attributes and textures of life-worlds in different regions of the globe. Within each of these forms of life imposed by different climatic conditions, there is an almost perfect impartiality and equality. Indeed, it would be most peculiar to suggest the opposite, namely that the impact of climate is somehow stratified and affects say the level of climatic energy of individuals depending on their social standing, their wealth or their political influence. On the contrary, the benefits and the costs associated with climate and therefore the destinies due to climate are almost always distributed without regard of those social and cultural factors social scientists otherwise would want to invoke as agents of social change, the identities of individuals, social mobility and inequality. Climate does not discriminate. The apparent lack of any selective, unmediated appropriation of climate in mentalities, its direct manifestation in cultural forms and social structures make climate determinism a highly unrealistic description of the interaction between nature and society.

1.6.3 *The stability of climate and conduct*

A further dubious element in the equation advanced by climate determinists concerns the often unacknowledged but evident stability and lack of fragility of social conduct. Climate not only does not discriminate; it also lacks for the most part any dynamic character and therefore the ability to insure anything but extremely stable life worlds. A steady and robust climate produces only static and repetitive consequences. Huntington does not entirely rule out the possibility of "phases of a long climatic cycle." In his early as well as in his last major work, he invokes the notion of long phases of climatic change in order to account for the shift in the fortune of regions and nations in the course of recorded history. For example, Huntington (1945:343) attributes the "Dark Ages" and the "Revival of Learning" in Europe to such a change in climatic conditions, more specifically, the prevalence of storms:

> *The Dark Ages and the Revival of Learning occurred at opposite phases of a long climatic cycle. Storminess apparently reaches a low ebb in the Dark Ages but an abundance and violence in the fourteenth century. These two periods were likewise times of psychological contrast. The Dark Ages were characterized by widespread depression of mental activity, whereas the Revival of Learning ushered in a period of alertness and hope.*

In his early work on climate and human affairs, for example, in his books *The Pulse of Asia* (1907) and *Palestine and Its Transformation* (1911) — both are narrative accounts of his travels in Central Asia in the years 1905–1906 and in the Middle East.[34] Huntington stresses climatic change, pulsations, periodicities and cycles both in historical and geological times, or short-term and long-term variations. He concedes that to him "who has devoted years to this particular line of study, they probably appear more important than they really are" (Huntington, 1913:222). He was convinced terrestrial climate changes are mainly due to fluctuations in the heat of the sun. In the final chapter of his book on Asia, Huntington (1907:359) summarizes the lessons of his observations concluding that "during historic times, climate, the most important factor in that environment [of Central Asia], has been subject to notable changes … it appears that the changes of climate have caused corresponding changes not only in the distribution of man, but in his occupation, habits, and even character."[35] Despite the caution and reservation Huntington himself issues, he quotes himself in the same 1913 essay and maintains that the rise and fall of civilizations occurs in close correspondence with favorable or unfavorable conditions of climate: "In the regions occupied by the ancient empires of Eurasia and north Africa, unfavorable changes of climate have been the cause of depopulation, war, migration, the overthrow of dynasties, and the decay of civilization; while favorable changes have made it possible for nations to expand, grow strong, and develop the arts and sciences" (Huntington, 1911:251).

Huntington's observations about the facts of climate change did not go uncontested. One of the first and prompt critics of his general thesis about the efficacy of climate vs. cultural (mental) factors, the historian A.T. Olmstead (1912), not only challenges his conclusions about the role of climate in the history of the Middle East but also the very assertion that these regions have been subjected to any significant change in climate in historical times that supposedly explain the fate of Middle East societies and fortune over time (e.g., Olmstead, 1912:166).

---

[34] *The Pulse of Asia* is one of the most reviewed geography books written by an American author in the early years of the 20th century.

[35] In a retrospective note that may be found in the copy of the second edition of *The Pulse of Asia* located in the library of the American Geographical Society, Huntington recalls that his "dominant motive in writing *The Pulse of Asia* was the hope that it would have a profound influence upon the course of human thought. I believed that in 'pulsations' of climate I discovered a key which would unlock some of the great mysteries of history" (cf. Martin, 1973:68).

But as we have already indicated the professional concern in both geography and climatology in the 1920s moved away from climatic change and increasingly stressed climatic stability in historical times. In the case of Huntington, he changes time horizons and becomes more concerned with the impact of what are actually weather patterns on human activities, for example, he examines rather short cycles in weather patterns, storms, days of great humidity etc. In the work of Huntington, attention to stable robust features of climate are liberally interwoven with comments about periodicities, long and short cycles and weather fluctuations. The attraction of such a liberal mixing, of extending and then collapsing the time horizon is of course that it makes any concerted effort to amass counterevidence very difficult if not impossible. Switching among time horizons becomes an effective strategy toward the immunization of the basic argument about the efficacy of climate.

Nonetheless, one of the frequent criticisms leveled against earlier climate determinists, for example, against the work of the philosophers of the French enlightenment concerned their assumption that climate, apart from the succession of seasons, was essentially stable.

1.6.4 *The dichotomous nature of climate*

One of the remarkable features of climate within climate determinism is its all or nothing quality; that is, climate determinism has the tendency to explicate the consequences in dichotomous categories. As a result, specific climatic conditions are, for example, either stimulating or its exact opposite, namely unstimulating, reflected in the diminished energy its inhabitants display — as the contrast between the climate of the State of New York and the State of Hawaii demonstrates according to Huntington (1945:390–391). Under stimulating conditions, such "matters as serious reading, inventions, new projects, and the promotion of education, health, and good government" get far more attention than in less stimulating climatic regions of the world. Although the kind of activities just enumerated are not completely absent, "they proceed more slowly than among people of similar ability, character, and training in more stimulating climates" and they tend to be "led by people who frequently go to the more bracing climates for education, recuperation, and stimulus" (Huntington, 1945:391–392).

Part of the one-dimensional analysis of climate in human affairs among climate determinists is also the uncanny way in which their analysis of how nature or climate works corresponds to their own opinions about humans and human society. Climate used in this way affirms that there cannot be an analytical reference to "climate in itself." Climate acquires its meaning in a particular context. One therefore is not only justified but forced to refer to the social construct of climate. What is the hidden model of climate in climate determinism?

1.7 The restriction of the range of social science discourse

As is well known but also widely supported, mainstream social science eliminated from consideration any perspective that made reference to natural forces as explanatory variables.[36] And it did so, as one should emphasize, for good reasons (cf. Grundmann and Stehr, 1997). As a result, social science discourse for the most part also has been successful in avoiding the seductive simplicity of most forms of technological, economic and biological determinism. Thus, the history of the social sciences in this century can also be written as a struggle against social Darwinism, racism, climate determinism and, to a great extent, socio-biology. Mainstream social science has succeeded in restricting its discourse to *sui generis* processes, such as social, political, economic or cultural. The basic problem for social theorists became how social order is possible. Any material or ecological conditions for the possibility of social order are treated as unproblematic or assigned by way of a division of intellectual labor to other academic disciplines.[37] Using a triad Werner Sombart employs, it is culture, technology and social structure that determines the foundation of social order. The social scientific perspective that now dominates is fundamentally opposed to the liberal mixture of explanatory dimensions one still encounters in the writings by climate determinists of this century. The fact that climate determinism continued to be practiced well into this century indicates that mainstream social science never fully succeeded in cleansing itself of inopportune intellectual perspectives, however

[36] As an early survey of proper social science conceptions by Floyd N. House (1929:16), therefore puts it with respect to climatic factors: "Questions of the sort with which Hippocrates and Ibn Khaldun concerned themselves are today regarded as the province of the physiologist." The ascent of the theoretical paradigms now taken for granted did not occur in tandem in social science disciplines; as a matter of fact geography is one of the exceptions; the vigorous environmental determinism in the early decades of this century in geography, now "often treated as part of geography's distant and shameful past" (Frenkel, 1992:146) is a case in poignant example.

[37] For classical social theorists, societal adaptation to environmental conditions surely was not the problem. The opposite appears to be self-evident for classic theory; Karl Marx (1974:517) and others were impressed and fascinated by the evident progress in the material capacity of distancing society from the constraints of nature: "The productive forces of mankind are immeasurable. The productiveness of the soil can be increased to infinity through the application of capital, work, and science."

much these perspectives were ostracized. Less radical attempts to alert social scientists to adaptive constraints and ecological dimensions, for example, as part of the human ecology perspective remained marginal within social science discourse.

The social sciences not only deliberately discarded references to physical, biological and generally environmental factors because they aspired to establish their own disciplinary, professional and academic identity firmly based on the definition of a subject matter that transcended that of the natural sciences; the social sciences also, for the most part, shared in certain ideological or moral assumptions related particularly to the notion of modernity and progress which incorporated the conviction that the march toward modern societies and desirable living conditions included an extensive emancipation from the immediate effect and dependence on environmental conditions. The liberation from (reductionist) naturalism is therefore a version of social emancipation.

The success social scientists generally have enjoyed in discarding and dismissing any reference to natural processes except in the vaguest sense of an insignificant background noise has been supported for decades by the view prevalent in natural science that nature exists in a state of equilibrium and permanence. Climate as an inert and essentially steady phenomenon can therefore easily be abandoned as a relevant dimension in social evolution, especially at a time of otherwise massive dramatic and often abrupt economic, political and social transformations around the world.

But now that impact of society on nature and, but less so, of "nature" on society are at the forefront of many discussions in science and politics; as a result, social science discourse is forced to re-examine its own relations to nature. Moreover, in elements of natural science discourse, the concept of "nature" increasingly is losing its static character and closed system attributes; it is depicted as mutable, dynamic as well as subject to human interference. Thus, the decades nature occupied a slum dwelling within social science discourse perhaps are numbered.[38] But most importantly, now that environmental factors are not merely a matter from which societies successfully distance themselves, considerations in social science discourse of climate matters, for example, acquire a new relevance.[39]

---

[38] The notion that nature is neither changeless nor cyclical did not of course emerge in the last few years but took decades to develop and has many intellectual parents as well as social developments that aided its development.

[39] The discovery of a possible reversal in the successful distancing of society from natural constraints is not a disclosure that could be expected to be made within social science today. It is a discovery that originates in models, images, concepts and research programmes in the natural sciences. But that does not mean that these issues should remain the exclusive domain of natural science discourse.

Also, now that the "evolution" of modern societies appears to have lost is immediately visible direction and drama, perhaps is even directionless, reference to natural processes and the impact or threat they are said to pose become a more credible perspective. However, the central task is to secure a sense of nature and climate in social science (as well as natural) discourse that transcends the intellectual traps liberally invited or perpetuated by modern climate determinism.

In short, we need to reconstitute the notion of nature in social science discourse. However, we have to avoid, on the one hand, the pitfalls of any (reductionist) naturalistic determinism including of course climate determinism and, on the other hand, remain satisfied with the mere introduction of the *topic* of the environment into social science discourse. Environmental sociology, for example, is the initial as well as most sustained effort in recent years to re-introduce environmental conditions into social science discourse. But for the most part, it is a plea to incorporate ecological topics into social theory thereby recognizing that society affects the environment. The environment continues to be located externally to society. Environmental sociology constitutes the environment within sociological discourse as a social problem analogous to many other and more traditional social problems such as deviant behavior, divorce and unemployment. As a result, environmental sociology has not succeeded in changing the paradigmatic relation of society and nature in social science discourse (cf. van den Daele, 1992). In addition to environmental sociology, there are other emerging efforts that propose a reconciliation of nature and society in social science discourse. One could refer to Bruno Latour's (e.g., 1993) heroic program to abandon the dualism of nature and society, the diverse work of feminist eco-sociology or those of neo-Marxist thinkers (e.g., Gorz, [1991] 1994). Our proposal stresses the need to *discover new phenomena* as the precondition for resisting the appeal of either naturalism or concepts that rely on a purely constructivist perspective. What is needed is the discovery that the "ecological deficit" in social theory extends primarily toward ways of incorporating "nature" into social science phenomena.

1.8 Climate matters

Everyone knows climate is important especially the extremes.

William F. Ogburn, 1943:785

Ellsworth Huntington ([1915] 1924:403) concludes his best-known work *Climate and Civilization* with what he declares is a farfetched warning. Huntington's prediction about the horrifying social, political and economic consequences of global climate change for the state of the world must have been for his contemporaries a thoroughly frightening scenario: "In a thousand years ... no highly favorable region may exist upon the globe, and the human race may be thrown back into the dull, lethargic state of our present tropical races."[40] Even discounting the possible and radical descent of advanced civilizations into a backward state of tropical societies, the prospects are clearly dismal, as Huntington concludes, because changes in the location of the regions around the globe with the highest "climatic energy and the consequent rise of new powers and the decline of those now dominant may throw the world into a chaos far worse than that of the Dark Ages. Races of low mental caliber may be stimulated to most pernicious activity, while those of high capacity may not have energy to withstand their more barbarous neighbors."

To move from Huntington's essentialist idea that climate works in such a definitive and context-insensitive manner to what we consider a more realistic notion that climate matters requires, first of all, the firm refusal to succumb to the seductive simplicity of climatic determinism and its fatalistic utopias. Although Huntington examined the "progress of civilizations" in relation to the natural environment and thereby anticipated or preceded many contemporary voices that demand such an inclusive perspective,[41] he actually did considerable damage, as we have tried to document, to a perspective that begins to reconcile the separation and alienation of nature and society in social science discourse.

Nature is no longer viewed as a regular, static entity and therefore climate is no longer seen as resting in a state of fixed equilibrium. Such a fundamental re-invention of nature should also have significant effects on the ways in which it might be re-introduced into social science discourse. We need to find a way of conceptualizing climate, for example, as a social construct that is not only a figment of our imagination and that does not merely refer to climate as "impacting" society.[42] But how can one conceptualize "climate'" in such a manner? That society is imprinted into nature is hardly controversial anymore because nature, as we

[40] If one describes Huntington's scenario as a "negative utopia" triggered, as it were, by massive climate changes, then his description of the societal consequences does not differ much in their characteristics from those found in more recent discussions about the potential effects of rapidly enhanced concentrations of greenhouse gases in the atmosphere.

[41] It therefore is perhaps worth noting that Ellsworth Huntington was a founding member of the Ecological Society of America.

[42] If one chooses to reject the idea that climate is also real, even offers resistance that is institutionalized in society and maintains instead that climate is merely a social construction, "the objectivity of nature and the objectivity of the ecological problem would vanish in a constructivist fog. We would

know and encounter it today, is in fact mainly a social construct. How does the natural climate demonstrate its "reality" in society and for social conduct? And how, for that matter, is nature generally inscribed and embedded into the fabric of society and thereby reflects the ways in which we comprehend natural processes in everyday life?

In a most general sense, we want to propose that natural and social processes are mainly imprinted or inscribed into the *boundary conditions* of each other, that is of nature and society. In the case of society, we would suggest that although it is significantly shaped by historical or selective constructions, our understanding of and encounter with climate is in important ways affected by, resonates with, and is shaped by "extreme" climatic responses — that at times may well be the consequence of human interventions into global climate processes.[43] The inscription of climate into society is never obvious and transparent but requires interpretation. Interpretations of how climate works on society are not completely arbitrary and divorced from the capacity and capability of climate to leave its imprints on society. We want to argue that the inscriptions climate achieves in society mainly operate via its extremes. What is apprehended as an extreme is by no means obvious. It is subject to different readings.

The evident fascination and even rapture with extremes of all descriptions among the public and the media in our times is well known, but probably not a novel response. Varied, even "ritualistic" cultural responses to extremes ultimately display and often celebrate the familiar, namely, "normal" patterns as well as homeostatic processes. Extremes constitute a "crisis" and are apprehended as temporarily disturbed equilibria. Using a term coined by William James, extremes are "coercive facts." It is in this sense that climate extremes violate taken-for-granted and trusted conceptions and observations about climate (Stehr, 1997). Although such extremes themselves likely are not interpreted as static over time, what is experienced as climate extremes are taken to be anomalies and disappointments. Climate extremes remind us of the reality hidden behind the surface of the social climate construct. Climate extremes offer and manifest the resistance of natural reality. As such they become imprinted into the social climate construct. They allow for the possibility of observing, categorizing and criticizing our observations about climate. In order to observe our observations about climate and its

then be dealing not with real risks, but with a 'construction' of crisis and not with real risks, but with mere perceptions of risks" (van den Daele, 1992:532).

[43] This assertion about the significance of extremes should not be misunderstood to represent a kind of ontological thesis rejecting any kind of "gradualism" (e.g., in natural history, see Gould, 1980:226) but as an empirical hypothesis about the practical ways in which nature becomes pertinent for society.

effects on society, we have to step back, we need to be forced to leave accepted interpretations or constructs. Climate *matters* as an event and a mechanism that precisely accomplishes this feat.[44]

That society has in the past responded to climate extremes that become imprinted into social action and continues to do so, can be shown easily because climate *extremes* are institutionalized (or inscribed as we also have called it) in society. For instance, they are inscribed in the form of a wide variety of myths, ideologies, stories (including more or less elaborated narratives of nature in everyday life), technologies, regulations, organizations, etc. An obvious as well as stable and powerful example is protective dikes erected at both rivers and oceans as well as the laws and regulations that govern their construction, maintenance and use. In much the same way, the evolution of shelter, clothing and nutrition is to some extent an inscription of climate extremes into the social fabric. Climatic extremes are engraved and objectified in the construction, maintenance and utilization of many of the modern means of transportation. Modern instruments of transportation are not only utilized to link open spaces with each other and carry commodities, information and humans but they constitute artifacts that are responses to climate, especially climatic extremes. In a way, means of transportation are portraits of and embody social encounters with climate. Of course, importantly such encounters manifest themselves in efforts to exclude or, to draw boundaries of exclusion for climatic extremes. Transportation takes place in familiar spaces, artificial zones and fairly tight enclosures that keep out undesirable climatic conditions. Nonetheless, engraved into the enclosure are climatic conditions or, nature that is not "our" nature from which we desire to withdraw. The greater the distance such artifacts have to travel the greater the likelihood that climatic extremes are inscribed into the construction of the object. As time and distance become increasingly irrelevant to social and economic life, the greater the influence of extremes on such construction of such artifacts. Paradoxically, as these extremes are built into the object they tend to vanish from view and certainly from direct experience and encounters.[45]

---

[44] In historical times, that is, climate matters but not as manifest, recorded or apprehended in form of gradual changes since these secular variations in meteorological phenomena such as temperature changes — that certainly are documented constituting reliable observational data — tend to occur in a very narrow range. The narrowness of the range of secular variations "and the autonomy of human phenomena which coincide with them in time, make it impossible for the present to conclude," as Le Roy Ladurie ([1967] 1988:275) stresses in his study of the interaction between climate and history since the year 1000, that "there is any causal link between them."

[45] Therefore, it is with justification that Marston Bates (1952:120) makes the following case against Huntington's thesis about the rise of civilizations, or, better, puts it on its head, when he says, "the

Although nature manifest in climate processes may be institutionalized in society and take on moral qualities (as in "nature strikes back" for example) of which it otherwise appears to be deprived, the institutionalization of nature paradoxically converts climate into an almost invisible entity. The institutionalization of climate in society paradoxically means to distance society from climate and decrease the contingencies for society that may issue from climate. The successful fabrication of a decline in the contingencies that arise from (the natural) climate allows for an increase in the contingencies that come with the socio-cultural development of knowledge.

1.9 Conclusions

At the end of the last and the beginning of this century, proponents of social science discourse and sociological discourse in particular, now considered to be major classics of their disciplines, discovered that social phenomena are unique in important respects, for example, in terms of their unmatched complexity as well as their unique developmental patterns that both demand and require a clear and distinct separation in explanatory principles and methodological procedures from the already then very successful natural sciences. Indeed, one of the enduring qualities of classical social science discourse is its insistence that social phenomena constitute a reality *sui generis*. The real virtue of this notion stems not so much from any inherent opposition between phenomena and the logic of their development, as they relate to the evident ethical and political consequences of attempts to relinquish or discursively join both attributes. It is a matter of historical record that any naive effort in this regard leads to a victory of reductionist conceptions (see Grundmann and Stehr, 1997).

In the 18th century, which according to many contemporary historians of social science represents the era in which modern social science discourse originated, was an age in which the educated part of the population in France, Germany and England spent enormous intellectual energy to argue about the climatic determinants of the civilizational peculiarities of entire nations (relying, e.g., on works by Montaigne, *Essais*, Montesquieu, *Esprit des Lois*, Falconer, *Remarks on the Influence of Climate*). As a contemporary observer was prompted to point out there was an endless number of writers who ascribe supreme efficacy to climate. Although the discussion of the impact of climate on societies did not cease

western European environment, lauded by Huntington and his followers as ideal for the development of civilization, was an insurmountable obstacle to civilization until methods had been found for mitigating its effects."

abruptly in social science, it ultimately was discredited and, only fairly recently, vanished almost without any trace as a largely compromised and widely discredited line of inquiry. It therefore has become more common today to find it "amusing to think that the men of former times would not have been put out by ... climatic explanation, implicating as it does the heavens" (Braudel, [1979] 1992:51).

There are good reasons that account for the differentiation of cognitive agendas in science, chief among them the following:

- biological and cultural evolution are not identical,
- the natural environment of society is for the most part independent of human action,
- societies have succeeded in emancipating themselves from many environmental constraints.

Nonetheless, the ecosystem, refashioned to a lesser or greater extent by social action by way of appropriating its resources, remains a major material source and constraint for human conduct. More recently, it has become evident, mainly as the result of research in the natural sciences that the emancipation of social conduct from nature is by no means firm and final. As a result, a re-examination of the well-entrenched intellectual division of labor in science may be in order. But such a revision of the asymmetric division among domains of inquiry will have to demystify first and foremost the persistent claim of natural science discourse to be located upstream and up front of social science. We have attempted to show how steps may be taken in this direction by suggesting to move the issue of the impact of climate on social action away from the established notion that climate works to the idea that climate matters for social conduct.

Bibliography

Arnold, David (1996) *The Problem of Nature. Environment, Culture and European Expansion*. Oxford: Blackwell.

Barnes, Harry Elmer (1921) "The relation of geography to the writing and interpretation of history." *The Journal of Geography* 20: 321–337.

Bates, Marston (1952) *Where Winter Never Comes. A Study of Man and Nature in the Tropics*. New York: Charles Scribner's Sons.

Beck, R.A. (1993) "Climate, liberalism and intolerance." *Weather* 48: 63–64.

Bernhard, Thomas ([1983] 1988) *Der Untergeher*. Frankfurt am Main: Suhrkamp.

Bourdieu, Pierre ([1982] 1990) "Die Wissenschaftlichkeitsrhetorik: Beitrag zu einer Analyse des Montesquieu-Effekts," pp. 169–179 in Pierre Bourdieu, *Was heißt sprechen? Die Ökonomie des sprachlichen Tauschs*. Wien: Universitätsverlagsbuchhandlung W. Braumüller.

Braudel, Fernand ([1979] 1992) *The Structures of Everyday Life. The Limits of the Possible. Volume 1 of Civilization and Capitalism 15th–18th Century*. Berkeley: University of California Press.

Brückner, Eduard (1923) "Julius Hann," pp. 151–160 in Akademie der Wissenschaften in Wien, *Almanach für das Jahr 1922*. Wien: Hölder-Pichler-Tempsky.

Elias, Norbert (1987) "The retreat of sociologists into the present," pp. 150–172 in Volker Meja, Dieter Misgeld and Nico Stehr (eds.) *Modern German Sociology*. New York: Columbia University Press (now London: Routledge, 2022).

Fischer, David Hackett (1996) *The Great Wave. Price Revolutions and the Rhythm of History*. New York: Oxford University Press.

Fleming, James (1999) *Historical Perspectives on Climate Change*. New York: Oxford University Press.

Flohn, Hermann (1941) "Die Tätigkeit des Menschen als Klimafaktor." *Zeitschrift für Erdkunde* 9: 13–22.

Freeman, Thomas Walter (1967) *The Geographer's Craft*. Manchester: Manchester University Press.

Frenkel, Stephen (1992) "Geography, empire, and environmental determinism." *The Geographical Review* 82: 143–153.

Gould, Stephen J. (1980) *The Panda's Thumb: More Reflections on Natural History*. New York: W.W. Norton.

Gouldner, Alvin W. (1980) "Is amnesia in sociology discontinuity, and the problem of permeable boundaries in culture," paper presented at an international conference on *The Political Realization of Social Science Knowledge*," Institute for Advanced Studies, Vienna, June 18–20.

Gorz, André ([1991] 1994) *Capitalism, Socialism, Ecology*. London: Verso.

Gilfillan, S. Colum (1920) "The cold ward course of progress." *Political Science Quarterly* 35: 393–410.

Gilfillan, S. Colum ([1935] 1970) *The Sociology of Invention: An Essay in the Social Causes, Ways and Effects of Technic Invention*, especially as demonstrated historically in the author's *Inventing the Ship*. Cambridge: MIT Press.

Grint, Keith and Steve Woolgar (1997) *The Machine at Work. Technology, Work and Organisation*. Oxford: Polity Press.

Grundmann, Reiner and Nico Stehr (1997) "Klima und Gesellschaft, Soziologische Klassiker und Aussenseiter. Über Weber, Durkheim, Simmel und Sombart." *Soziale Welt* 47: 85–100.

Hann, Julius (1883) *Handbuch der Klimatologie*. Stuttgart: J. Engelhorn.

Hann, Julius (1903) *Handbook of Climatology. Part I: General Climatology*. New York: Macmillan.

Herder, Johann Gottfried (1794) *Ideen zur Philosophie der Menschheit. Zweiter Theil.* Carlsruhe: Christian Gottlieb Schmieder.

Hoheisel, Karl (1993) "Gottesbild and Klimazonen," pp. 127–140 in Ruprecht-Karls-Universität Heidelberg (ed.), *Studium Generale 1992*. Heidelberg: Heidelberger Verlagsanstalt.

House, Floyd N. (1929) *The Range of Social Theory. A Survey of the Development, Literature, Tendencies and Fundamental Problems of the Social Sciences*. New York: Henry Holt.

Hulme, Mike (2011) "Reducing the future to climate: A story of climate determinism and reductionism." *Osiris* 26: 245–266.

Huntington, Ellsworth

—— 1945 *Mainsprings of Civilization*. New York: John Wiley and Sons.

—— 1935 *The Goal of Eugenics*.

—— 1928 "Temperature and the fate of nations." *Harper's Magazine* 157: 361–368.

—— 1927a *The Human Habitat*. New York: Van Nostrand.

—— 1927b "Sociological relationships of climate and health," pp. 257–284 in Jerome Davis and Harry E. Barnes (eds.) *Readings in Sociology*. New York: D.C. Heath.

—— 1924a "Environment and racial character," pp. 281–299 in Malcolm Rutherford Thorpe (ed.), *Organic Adaptation to Environment*. New Haven: Yale University Press.

—— 1924b *The Character of Races as Influenced by Physical Environment, Natural Selection and Historical Development*. New York, London: C. Scribner's Sons, 1924.

—— 1916 "Climatic variations and economic cycles." *The Geographical Review* 1: 192–202.

—— [1915] 1924 *Civilization and Climate*. Third Edition, revised and rewritten with many new chapters. New Haven: Yale University Press.

—— 1914a "Climatic changes." *The Geographical Journal* XLIII: 293–313.

—— 1914b "The geographer and history." *The Geographical Journal* XLIII: 19–32.

—— 1913 "Changes of climate and history." *American Historical Review* 18: 213–232.

—— 1912 "Letter." *Bulletin of the American Geographical Society* 44: 440–447.

—— 1911 *Palestine and Its Transformations*. New York: Houghton, Mifflin & Co.

—— 1907 *The Pulse of Asia*. Boston: Houghton & Mifflin.

Huntington, Ellsworth and Samuel S. Visher (1922) *Climatic Changes. Their Nature and Causes*. New Haven, Connecticut: Yale University Press.

Huntington, Ellsworth and Sumner W. Cushing (1921) *Principles of Human Geography*. New York: John Wiley & Sons.

Khaldûn, Ibn ([1377] 1958) *The Muqaddimah. An Introduction to History*. Volume One. Princeton, New Jersey: Princeton University Press.

Latour, Bruno (1993) *We Have Never Been Modern*. Cambridge: Harvard University Press.

Leroy-Beaulieu, Anatole (1893) *Empire of the Tsars and the Russians*. Volume 1. New York: Putnam.

Le Roy Ladurie, Emmanuel ([1967] 1988) *Times of Feast, Times of Famine: A History of Climate Since the Year 1000*. New York: Farrar, Strauss and Giroux.

Livingstone, David N. (1991) "The moral discourse of climate: Historical considerations on race, place and virtue." *Journal of Historical Geography* 17: 413–434.

Livingstone, David N. (2011) "Environmental determinism," pp. 368–380 in John A. Agnew (ed.), *The Sage Handbook of Geographical Knowledge*. Sage: Los Angeles.

Martin, Geoffrey J. (1973) *Ellsworth Huntington. His Life and Thought*. Hamden, Connecticut: The Shoe String Press.

Marx, Karl (1974) *Die Frühschriften. Karl Marx and Friedrich Engels*. Collected Works, Vol. 1. Berlin: Dietz.

Massin, Benoit (1996) "From Virchow to Fischer: Physical anthropology and 'modern race theories' in Wilhelmine Germany," pp. 79–154 in George W. Stocking Jr. (ed.), *Volksgeist as Method and Ethic*. Essays on Boasian Ethnography and the German Anthropological Tradition. History of Anthropology, Volume 8. Madison, Wisconsin: The University of Wisconsin Press.

Merton, Robert K. and M.F. Ashley-Montagu (1940) "Crime and the anthropologist." *American Anthropologist* 42: 384–408.

Mills, C. Wright (1959) *The Sociological Imagination*. New York: Oxford University Press.

Moore, Henry L. (1914) *Economic Cycles. Their Law and Cause*. New York: Macmillan.

Nordhaus, William D. (1994) "The ghosts of climate past and the specters of climate future," pp. 35–62 in Nakicenovic, Nordhaus, Richels, and Toth (eds.), *Integrative Assessment of Mitigation, Impact and Adaptation to Climate Change*. Laxenburg: IIASA.

Ogburn, William F. (1943) "Review of Clarence A. Mills, *Climate Makes Man*." *American Journal of Sociology* 48: 784–787.

Olmstead, Albert T. (1912) "Climate and history." *Journal of Geography* 163–168.

Semple, Ellen Churchill

—— 1931 *The Geography of the Mediterranean Region. Its Relation to Ancient History*. New York: Henry Holt and Company.

—— 1911 *Influences of Geographic Environment*, on the basis of Ratzel's system of anthropo-geography. New York: Holt, Rinehart and Winston.

Sombart, Werner (1931) "Die Entfaltung des modernen Kapitalismus," pp. 85–104 in Bernard Harms (ed.), *Kapital und Kapitalismus. Vorlesungen gehalten in der Deutschen Vereinigung für Staatswissenschaftliche Fortbildung*. Berlin: Reimar Hobbing.

Sorokin, Pitirim A. (1928) *Contemporary Sociological Theories*. New York: Harper & Brothers.

Spate, Oskar Hermann Khristian (1952) "Toynbee and Huntington: A study in determinism." *The Geographical Journal* 118: 406–428.

Spencer, Herbert (1887) *The Study of Sociology*. London: Kegan Paul, Trench & Co.

Starr, Kevin (1985) *Inventing the Dream. California through the Progressive Era*. New York: Oxford University Press.

Stehr, Nico (1997) "Trust and climate." *Climate Research* 8: 163–169. (1994) *Knowledge Societies*. London: Sage.

Stehr, Nico and Hans von Storch (1997) "Rückkehr des Klimadeterminismus?" *Merkur* 51: 560–562.

Thomas, Franklin (1925) *The Environmental Basis of Society. A Study in the History of Sociological Theory*. New York:

Thomas, William I. (1909) *Source Book for Social Origins. Ethnological Materials, Psychological Standpoints Classified and Annotated Bibliographies for the Interpretation of Savage Societies*. Boston: Richard G. Badger.

Tolleefson, Jeff (2016) "The hostile ocean that slowed climate change." *Nature* 539 (November): 346–348.

Toynbee, Arnold J. (1973) "Foreword," pp. ix–x in Geoffrey J. Martin, *Ellsworth Huntington. His Life and Thought*. Hamden, Connecticut: The Shoe String Press.

Van den Daele, Wolfgang (1992) "Concepts of nature in modern societies and nature as a theme in sociology," pp. 526–560 in Meinolf Dierkes and Bernd Biervert (eds.) *European Social Science in Transition. Assessment and Outlook*. Frankfurt am Main: Campus.

Virchow, Rudolf ([1885] 1922) "Über Akklimatisation," pp. 214–239 in Karl Sudhoff (Hrsg.), *Rudolf Virchow und die deutschen Naturforscherversammlungen*. Leipzig: Akademische Verlagsanstalt.

Wallerstein, Immanuel *et al.* (1996) *Open the Social Sciences. Report of the Gulbenkian Commission on the Restructuring of the Social Sciences*. Stanford, California: Stanford University Press.

Weber, Max ([1909] 1922) "'Energetische' Kulturtheorien," pp. 376–402 in Max Weber (Hrsg.), *Gesammelte Aufsätze zur Wissenschaftslehre*. Tübingen: J.B.C. Mohr (Paul Siebeck).

Weinstein, Jay and Nico Stehr (1999) "The power of knowledge: Race science, race policy, and the holocaust." *Social Epistemology* 13: 3–36.

Appendix: The efficacy of climate. An inventory[46]

Alcoholism (Semple, 1911:626).

Arrests (Huntington, 1945:363–364).

Asiatic handicap ("In Europe and especially Asia the value of the climate as an aid to civilization declines quite steadily eastward," Huntington, 1945:385).

Attitudes ("People feel growingly optimistic in the spring and still more so in the autumn," Huntington, 1945:318).

Business activities and cycles (In almost every advanced country has sharp seasonal variations in its occupations, wages, trades, transportation, bank clearings, and other phases of business," Huntington, 1945:312).

Capacity for work ("Differences in health indicate corresponding differences in inclination to work, as well as in actual capacity to work. Vigorous people prefer to work rather than sit idle. The will to work beyond the required limits is extremely important in crisis, such as war, flood or other disaster. It is one of the main factors in leading people to make inventions, explore new lands, carry out scientific experiments, initiate reforms, and produce works of art, literature, and music," Huntington, 1945:238).

Circulation of books (Huntington, 1945:610).

Civilizations, distribution of ("As the Tropics have been the cradle of humanity, the Temperate Zone has been the cradle and school of civilization. Here Nature has given much by withholding much," Semple, 1911:635; Fig. 86 "Map of Civilization" on p. 256 in Huntington and Cushing, 1921:256; "The distribution of

[46]As already noted, not unlike political discourse, one of the systemic features of the narrative of climate determinism is its redundancy as if an assertion gains credibility by being repeated and repeated. For example, no attempt has been made therefore to list every instance in which specific assertions about the climate as a cause are advanced in the writings of Ellsworth Huntington. Huntington's observations about the importance of climate for the emergence of slavery in the Southern United States, for instance, can be found throughout his various writings covering a period of more than four decades. The inventory tries to be complete in a different sense, insofar as most of the alleged causal linkages between climate and social processes have been collected in the appendix.

civilization throughout the world has always depended closely upon climate," Huntington, 1927:165; "By encouraging one type of social organization and discouraging another, climate has great influence upon the development of civilization," Huntington, 1945:276).

Civil war (In the United States: "In all these respects climatic contrasts paved the way for civil war," Huntington, 1945:280).

Cleanliness ("The climate itself may also be largely responsible for the lack of cleanliness [in this case, among Icelanders]. So far as I am aware, this lack prevails among every people who live in a cool, moist climate where the water is always cold and where animals are the chief means of support. ... The cleanest people in the world are the inhabitants of warm, moist countries, where the state of culture requires clothing, and where there is plenty of water," Huntington, 1924b:289).

Commerce (The decline and rise of commercial activities as dependent on climate, e.g., Huntington, 1924b:300).

Communication (As dependent on favorable climatic conditions, e.g., Huntington, 1924b:300).

Crime (Huntington, 1945:365–367).

Cultural development ("Climate ... helps to influence the rate and the limit of cultural development. It determines in part the local supply of raw material with which man has to work, and hence the majority of his secondary activities, except where these are expended on mineral resources. It decides the character of his food, clothing, and dwelling, and ultimately of his civilization," Semple, 1911:609; "the North Temperate Zone is preëminantly the cultural zone of the earth," Semple, 1911:634; "Cultural variations from season to season seem to be intimately connected with physiological conditions that manifest themselves in reproduction and in rate of work," Huntington, 1945:319).

Cultural patterns ("Cultural habits rarely survive and thrive if they are actively in opposition to the demands of the physical environment," Huntington, 1945:319).

Cycles of activity ("Annual cycle of mental activity, which is especially clear in the circulation of serious books," Huntington, 1945:610).

Decline or decadence of civilizations ("The question has been repeatedly raised as to whether there have been changes in climate in historical times, especially rainfall fluctuations, sufficient to explain the decline and fall of the Roman Empire and the decadence of civilization, by reason of which large sections of the Mediterranean lands, once thriving and populous, have become depopulated or impoverished. Arguments supporting this position have been advanced chiefly by historians, archeologists, and other incompetent authorities not concerned with climatology. The majority of competent authorities have reached a contrary conclusion ... Ellsworth Huntington attributed the decline of Palestine, Syria, Asia Minor, Greece and Italy to the same cause, but his arguments have been questioned both by historians and climatologists," Semple, 1931:99–100).

Degeneration (As the climate becomes unfavorable — as in cold and stormy in Iceland, e.g., Huntington, 1924b:293).

Diseases (The impact of climate on health is stressed by many climate determinists, even though it may only be in a kind of superficial and less consequential fashion — see below — than the stronger assertion that infectious diseases of one sort or the other are either promoted or repressed by climatic conditions: "Climate undoubtedly modifies many physiological processes in individuals and peoples, affects their immunity from certain classes of diseases and their susceptibility to other," Semple, 1911:608).

Dishonesty (see Stupidity).

Economic cycles ("The rhythm in the activity of economic life, the alternation of buoyant, purposeful expansion with aimless depression, is caused by the rhythm of the yield per acre of the crops; while the rhythm in the production of the crops is, in turn, caused by the cyclical changes in the amount of rainfall. The law of the cycles of rainfall is the law of the cycles of crops and the law of economic cycles," Moore, 1914:135).

Economic prosperity and development ("Economic prosperity and general well-being are distributed according to much the same geographical pattern as social welfare," Huntington, 1945:232).

Efficiency ("Extremes both of heat and cold reduce the density of population, the scale and efficiency of economic enterprises," Semple, 1911:611).

Elites (see Inequality).

Energy and progress (Fig. 85 on p. 255 in Huntington and Cushing, 1921, "Map of Climatic Energy" shows how "human energy would be distributed if it depended wholly on climate"; the map sums up the "combined effects of temperature, humidity, seasons and storms upon health and energy," Huntington, 1927:145; "the energy and progress of the world's leading countries is due to the constant repetition of the physiological stimulus which comes with the changing seasons," Huntington, 1945:319).

Fertility (Virchow, [1885] 1922:231).

First-class factories (Gilfillan, [1935] 1970:49; G. calls climate the most *fundamental* among the variables he examined).

Health. One of the more frequently cited effects of the climate is that on health. ("The climate of Iceland is not only healthful but stimulating," Huntington, 1924b:289; "the geographical distribution of health and vigor depends largely on the combined effect of climate and cultural conditions," Huntington, 1945:240; "in the United States infants conceived in the fall and born in the summer are especially numerous and have the lowest percentage of congenital defects," Huntington, 1945:319; "the resistance of infants ... to digestive diseases apparently varies according to their age in a way that suggests an innate adaptation to a particular kind of climate. The peculiar ability of people, especially women, in the reproductive ages of life to resist disease during the late winter suggests the same thing," Huntington, 1945:610).

History ("The greatest events of universal history and especially the greatest historical developments belong to the North Temperate Zone," Semple, 1911:611; "where man has remained in the Tropics, with few exceptions he has suffered arrested development," Semple, 1911:635).

Homicide ("Homicide shows a significant relation to temperature both geographically and seasonally ... seasonally as well as geographically, the rates increase from cooler to warmer weather ... warm weather apparently is associated with lowered self-control. It also makes people feel disinclined toward steady effort. Lack of self-control is a primary factor in the failure of public sentiment to express itself in observance of law," Huntington, 1945:232).

Immorality (see Stupidity).

Inequality ("The old South distinguished sharply between aristocrats and 'poor whites', as well as between whites in general and Negroes. This distinction of classes was in strong contrast to the relative democracy which prevailed in the North, where the squire might care for his own horse, cow and garden. When slavery disappears, a system of tenancy almost invariably grows up in regions where differences in ability to manage people and property are especially important in comparison with the ability to do manual work.").

Insanity ("At that time [June] the physical stimulus which merely leads to health and increased powers of reproduction among normal people apparently overestimates those who are poorly poised, weak of will, oversexed, or otherwise abnormal," Huntington, 1945:365).

Intelligence ("People of high latitudes are, on the whole, more intellectual than those of low latitudes," Huntington, 1945:367).

Inventions (Huntington, 1945:391).

Life expectancy (Huntington, 1945:610).

Mental activity ("Among European races physical activity appears to be the greatest when the temperature averages not far from 65°F, whereas mental activity seems to be greatest at a lower temperature, averaging perhaps 40°," Huntington, 1924b:290; in addition, climate variability stimulates mental activity, e.g., Huntington, 1924b:290).

Migration ("The acclimatization of tropical people in temperate regions will never be an equation of widespread importance. … [The Negroes'] concentration in the 'black belt', where they find the heat and moisture in which they thrive, and their climatically conditioned exclusion from the more northern states are matters of local significance. Economic and social retardation have kept the hot belt relatively underpopulated," Semple, 1911:625–626; "the people in poorer climates are practically certain to have poorer health and less energy than others. The population as a whole is likely to be less prosperous, so that education and contact with other people are less prevalent. Moreover, under such circumstances there is a strong tendency for the more able people to leave the poorer environment,"

Huntington, 1927:162; "Climatic conditions begin to mold and select the migrants to the new environment," Huntington, 1927:165).

Mortality ("Bodily temperature rises [in the Torrid Zone], while susceptibility to disease and rate of mortality show an increase ominous for white colonization," Semple, 1911:626).

National character (Huntington, 1945:303).

Patent productivity ("An isoplethic [or 'contour'] map I have made, of American patent productivity per capita, shows a heavy concentration in the narrow belt of best climate, near the 50°F isotherm, from Chicago to Philadelphia and Boston," Gilfillan, [1935] 1970:46).

Physical activity ("Physical vigor is basic in human progress. ... Vigor is needed in order that people may work hard without undue fatigue and have a reserve of strength in emergencies. It is especially important in promoting mental activity and clear thinking," Huntington, 1945:237; "Physical vigor is one of the main factors in the growth of civilization," Huntington, 1945:275; the "optimum temperature depends upon the conditions under which man took the evolutionary steps which gave him his present adjustment to climate," Huntington, 1945:273; "at temperatures above the optimum, fatigue is readily induced, the inclination to work diminishes, and the easiest way to make oneself conformable is to do as little as possible. At temperatures below the optimum the inclination to work is stimulated, partly because bodily activities promote warmth, partly because there are many ways in which a moderate degree of inventiveness enables people to keep themselves warm artificially," Huntington, 1945:275).

Physiology ("The effects of a tropical climate are due to the intense heat, to its long duration without the respite conferred by a bracing winter season, and its combination with the high degree of humidity prevailing over most of the Torrid Zone. These are conditions that are advantageous to plant life, but hardly favorable to human development. They produce certain derangement in the physiological functions of heart, liver, kidneys and organs of reproduction," Semple, 1991:626).

Productivity (see Capacity for work; Energy and progress).

Profitability ("The climate makes certain occupations profitable, and others unprofitable," Huntington, 1927:165).

Progress ("A map of climate, or rather of climatic energy, as we may call it, resembles a map of progress far more closely than does a map of any other factor which may be a cause rather than a result of the distribution of progress," Huntington, 1927:140).

Prostitution and sexual extravagance ("Seem to reach a maximum in the hottest parts of the world, that is, in the dry parts of a belt located ten to thirty degrees from the equator," Huntington, 1945:296).[47]

Reading, serious (Huntington, 1945:391).

Religion ("Diversity of physical environment has also been effective in leading to religious differences, and among the environmental factors climate has been especially important," Huntington, 1945:281).[48]

Reproduction (The reproductive "cycle varies according to climate." In the northern United States and western Europe, the maximum of births normally occurs in March or April as a response to conceptions in June or July. Elsewhere the maximum tends to shift to earlier dates in hot climates and later ones where the climate is cold" (Huntington, 1945:273–274).

Revolutions ("In the world as a whole the tendency toward lack of self-control in politics, in sex relations, and in many other respects rises markedly in hot weather

[47] Huntington (1945:296) refers, in this context to Hellpach (without further specificity; however, in the bibliography, Willy Hellpach's 1911 Die geopsychischen Erscheinungen des Wetters, Klima und Landschaft in ihrem Einfluss auf das Seelenleben is listed) and quotes him as saying that "in Southern Italy sexual irregularities increase greatly when the sirocco is blowing. The people recognize this so well that offenses committed under such circumstances are in a measure condoned."

[48] Since religious belief systems are not merely other-worldly but from this world, early mythological and later more systematic religious belief systems always display certain environmental constraints with which their originators struggled, and they even tend to reflect or incorporate certain climatic conditions (cf. Hoheisel, 1993) but this is of course a far cry from maintaining an almost indiscriminate assertion that religious beliefs and practice are driven by climatic conditions. Moreover, as Hoheisel (1993:130) points out, available ethnographical information lack reliability and validity to clearly tie religious beliefs and practices to climatic conditions: "In any case, increasing spatial mobility and progressive liberation from natural constraints, for example, through trading and commerce at a distance, but above through the possibility of being able to build on traditions of very different origins, make it considerably more difficult to prove that beliefs or other religious doctrines are characterized by certain climatic conditions."

and in hot countries. This is not the only reason for the frequency of political revolutions in low latitudes, but it must play a part," Huntington, 1945:365).

Riots ("Weather as a promoter of riots has hitherto been neglected. Nevertheless, it seems to agree with the distribution of riots [in India]"; "it is noteworthy that in the United States Negro riots occur most often in unusually hot weather," Huntington, 1945:362,364).

Self-control (Climatic "extremes weaken the power of self-control," Huntington, [1915] 1924:404; there is "evidence that dry weather, especially when hot, is associated with a decline in self-control," Huntington, 1945:296).

Sexual offenses (Huntington, 1945:365).

Slavery ("It was not only the enervating heat and moisture of the Southern States, but also the large extent of their fertile area which necessitated slave labor, introduced the plantation system, and resulted in the whole aristocratic organization of society of the South," Semple, 1911:622; "Slavery failed to flourish in the North not because of any moral objection to it, for the most godly Puritans held slaves, but because the climate made it unprofitable," Huntington, [1915] 1924:41; "The suppression of slavery in the North was not due chiefly to moral conviction. That arose after long experience had shown that slavery did not pay in a cool climate ... the combination of good food, stimulating climate, and northern type of culture made the white northerners so energetic that it irked them to wait for slow-moving Africans," Huntington, 1945:279).

Scientific research ("... the world's scientific research and other intellectual activities, as well as its financial, commercial, industrial and political control are more and more becoming concentrated in the few limited regions where the climate is most healthful and stimulating," Huntington, 1927:160).

Social ideals ("The difference in inclination toward work had much to do with the development of diverse social ideals in these parts of the United States. In the North, the successful family was the one where everybody worked hard as well as intelligently. Hard work became the supreme virtue, as it is to this day in spite of other tendencies. In the South, the successful ante-bellum family was one that eschewed physical labor and at the same time got a good living. This system favored slavery and attached a social stigma to work with the hands. An aristocratic society was almost inevitable because the mental ability to get a good living through slave labor is more limited than the physical ability which was so important in the North," Huntington, 1945:280).

Social systems ("In the United States, we see a social system closely in accord with the stimulating seasonal changes and storms which characterize the culture. We also see that the combined effect of the climate and the social system is so strong that children are especially active here, manufacturing and other forms of business forge ahead with a zest rarely seen elsewhere," Huntington, 1945:341).

Stupidity ("The climate of many countries seems to be one of the great reasons why idleness, dishonesty, immorality, stupidity, and weakness of will still prevail," Huntington, [1915] 1924:411).

Suicide ("In 1922, four California cities led the list of suicides. ... Possibly these facts may be connected with the constant stimulation of the favorable temperature and the lack of relaxation through the variations from season to season and from day to day, although other factors must also play a part. The people of California may perhaps be likened to horses which are urged to the limit to that some of them become unduly tired and break down," Huntington, [1915] 1924:225; Huntington, 1945:365).

Superstition (e.g., Huntington, 1924b:297).

Temperament ("The northern peoples of Europe are energetic, provident, serious, thoughtful rather than emotional, cautious rather than impulsive. The southerners of the sub-tropical Mediterranean basin are easy-going, improvident except under pressing necessity, gay, emotional, imaginative, all qualities which among Negroes of the equatorial belt degenerate into grave racial faults," Semple, 1911:620).

Tempo of social change ("The compression of climatic differences into a small area enlivens and accentuates the process of historical development," Semple, 1911:618).

Thinking (see Mental activity).

Thrift ("The necessity of preparing shelter, clothing, and fuel as means of combating the cold and moisture of winter tends to promote a social system which places high value on foresight and thrift," Huntington, 1945:277).

Unrest and violence (see Riots).

Wages ("The low cost of living keep down [the] wages, so that the laborer ... is poorly paid [in southern countries and regions] ... The laborer of the north, owing to his providence and larger profits, which render small economies possible, is constantly recruited into the class of capitalist," Semple, 1911:620–621).

Work attitudes ("A hot climate, especially if it is humid, makes people feel disinclined to work. This encourages the cleverer people to get a living with as little physical exertion as possible. Their example fosters the growth of a social system in which hard work is regarded as plebeian," Huntington, 1945:276; "the greatest social influence [of climate] is probably its effect on inclination to work," Huntington, 1945:282).

2. On the Power of Climate. Is Climate Determinism Just a History of Ideas or a Relevant Factor in Current Climate Policy?†

Abstract

Until the 1980s, climate dynamics were the focus of climate research, but since the 1990s, the focus has been on the threat of a "climate catastrophe" and climate change mitigation. The authors argue that this kind of research no longer calls only for natural scientists, but equally for social and cultural scientists. Our ideas about the dangers associated with climate change are only partially an expression of scientific knowledge, but to a considerable extent have their origins in pre-scientific and outdated forms of knowledge. Here, climate determinism plays a special role, which, although long since discredited in the social sciences, nevertheless plays an important, subliminal role in today's climate debate.

2.1 Introduction

Anyone who has read Emile Durkheim's *Suicide*[49] (first published in 1897) will be familiar with his classic methodological argument, which has become the paradigm of modern sociology, that apparently completely idiosyncratic individual actions are social phenomena or that their distribution cannot be attributed to physical or even cosmic causes.

Many fellow scientists of his time, on the other hand, were convinced that there was a causal relationship between climate or weather and the number of suicides. Durkheim's judgment is severe: "the facts must be in unusual agreement to require such a hypothesis [...] We must therefore seek the cause of the unequal inclination of peoples for suicide, not in the mysterious effects of climate but in the nature of this civilization, in the manner of its distribution among the different countries."

Where environmental determinism leaves off, the social sciences begin.

---

† This section originally appeared in Stehr, N. and H. von Storch: "Von der Macht des Klimas. Ist der Klimadeterminismus nur noch Ideengeschichte oder relevanter Faktor gegenwärtiger Klimapolitik?" (translated by the authors), *Gaia* 9: 187–195, 2000.

[49] Émile Durkheim: *Suicide*. Routledge, London (1952). The quote can be found on pp. 54–55.

Durkheim's work had a spectacular impact in the history of ideas of the social sciences. The separation between social and natural sciences is cemented and celebrated in them. On the other hand, Durkheim's efforts to radically overcome the fallacies of environmental determinism found remarkably little intellectual resonance in other social science disciplines at the time. The real scientific flowering of climate determinism and the in some respects related, but also competing, science of race came only at the beginning of the twentieth century.[50]

In the post-World War II era, the idea of the once-dominant climate determinism or geo-determinism appeared as a simple-minded, template view of the world. Among serious scientists, there was little intellectual incentive to further develop these paradigms, and the same was true for decision makers in business and politics.

In recent years, however, these fields of research have experienced a kind of renaissance, mostly as a rediscovery of an old way of thinking: In terms of the history of ideas, much of today's climate impact research is unadulterated climate determinism.[51] In recent years, this politically relevant line of research has all too often made use of the "dumb farmer" approach, according to which changing climate conditions encounter an unchanging social and economic reality and therefore become predictable in their consequences. Geodeterminism also still has supporters today. The American economic historian David Landes[52] argues that the time has come to put climatic and geographic factors in the context of comparative studies of the prosperity and poverty of nations as a dimension that conforms to reality. He regrets that Huntington's climate determinism has long discredited this approach, and adds, "geography also bothers many people because it is obviously and intrinsically unequal. There are places with better climates and worse climates from the point of view of comfort and health, and a lot of social scientists are reluctant to accept this. It really bothers them to see evidence of nature's 'unfairness'".[53]

Accordingly, there are several reasons to revisit the paradigm of classical scientific climate determinism. The aim is to prevent repetition of the excesses, misinterpretations and misunderstandable generalizations in a modern climate- and geo-reductionism, especially because the climate determinism is latent among

---

[50] J. Weinstein, N. Stehr: "The power of knowledge: Race science, race policy, and the Holocaust," *Social Epistemology* 13 (1999): 3–36.

[51] N. Stehr, H. von Storch: *"Rückkehr des Klimadeterminismus?"* *Merkur* 51 (1997): 560–562.

[52] D. Landes: *The Wealth and Poverty of Nations. Why Some Are So Rich and Some Are So Poor.* W.W. Norton, New York (1998).

[53] D. Landes: "Culture counts: Interview with David S. Landes," *Challenges* 41 (1998): 14–30, especially pp. 14–16.

natural scientists, but as a concept, it is almost forgotten, because it is easily dismissed by social and cultural scientists as an outdated episode in the history of ideas. On the other hand, the pitfalls of this paradigm must be analyzed in the light of the demands for a fundamental revision and overcoming of the deep cultural division of natural, social, and cultural sciences, which are legitimized especially by pressing environmental problems.[54] Finally, and this may be the most important aspect, climate determinism plays an active role in the formation of public, politically effective concepts about climate and climate change. Scientific knowledge about possible or even true climate change is transformed into politically effective knowledge, partly through the detour of climate impact research, partly through media appropriation, with the help of common everyday ideas.

In our essay, we attempt to capture the core messages of scientific climate determinism, particularly in the first half of the 20th century. We analyze the epistemological interests, scientific theoretical premises, and methodological procedures of climate determinism. We point out the historical context in which this tradition not only found considerable scientific resonance, but also social recognition, and deliberately sought to provide practical knowledge for the adherents of climate determinism.

We focus on the ideas and theories of two representatives of the modern "environmental determinism." Both scientists are well-known geographers of the first half of the 20th century: The American Ellsworth Huntington and the German Eduard Brückner. Both geographers exercised great intellectual influence; Huntington, in particular, had a significant impact on public opinion and the North American political class.

In Section 2, we discuss the ideological and historical background of climate determinism. In Section 3, we discuss Huntington's work in more detail. Our presentation of Brückner's ideas in Section 4 is shorter since we have already published extensively on Brückner elsewhere.[55] In Section 5, we deal with the separation of social and natural sciences as currently practiced. The continued existence of the so-called two scientific cultures can be understood in part as a reaction to the excesses of climate determinism. Today, it hinders a comprehensive, problem-oriented investigation of the relationship between society and nature.

[54] I. Wallerstein: *Open the Social Sciences. Report of the Gulbenkian Commission on the Restructuring of the Social Sciences.* Stanford University Press, Stanford, California (1996), especially p. 76.

[55] N. Stehr, H. von Storch, M. Flügel: "The 19th century discussion of climate variability and climate change: Analogies for present day debate?" *World Resources Review* 7 (1996): 589–604.
N. Stehr, H. von Storch (eds.): *Eduard Brückner — The Sources and Consequences of Climate Change and Climate Variability in Historical Times.* Kluwer, Dordrecht (2000).

In an outlook (Section 6), we emphasize the importance of interdisciplinary work, in which, however, the social sciences must not, as has often been the case, be degraded to stooges of the natural sciences.

2.2 The development of a school of thought

For centuries, scientists, intellectuals, humanists, philosophers, physicians, and certainly large segments of the population had little doubt about the extraordinary social and psychological effectiveness of climate.[56] The impact of climate on the physical and psychological characteristics and worldviews of people, both in their own societies and among the inhabitants of neighboring and more distant regions, was probably the most important factor and first discussed in more detail by Hippocrates of Cos (c. 460–377 B.C.) in his work "Air, Water and Place." A little later, Aristotle (384–322 B.C.) identified the climate as the cause of the superiority of the Greeks over the barbarians, thus confirming the typically expressed suspicion that their own climate was superior to that of foreign lands.[57]

Thinkers like Montaigne (1533–1592), Montesquieu (1689–1755), Herder (1744–1803) and Falconer (1744–1824) took the old theories to new heights, so that by the end of the 19th century climatic determinism was part of textbook[58] and encyclopedia knowledge.[59] Differences between peoples were reduced to climatic factors as a matter of course.

The development of climate determinism as an influential scientific school of thought in the social and natural sciences reached its peak in the first two decades of the 20th century. In studies of the influence of the natural environment on the course of human history, natural scientists, anthropologists, sociologists, physiologists and geographers developed an increasingly quantitative-empirical and therefore "objective" approach to the social and psychological significance of climate.

Some of the most memorable and convincing statements on the comprehensive, pernicious consequences of climate change were published during this period, although in the end they merely repeated centuries-old assertions and prejudices. Thus, the well-known geographer Semple begins her frequently cited

[56] C. Glacken: *Traces on the Rhodian Share*. University of California Press, Berkeley (1967).

[57] H. Bames: "The relation of geography to the writing and interpretation of history," *The Journal of Geography* 20 (1921): 321–337.

[58] F. Umlauff: *Das Luftmeer. Die Grundzüge der Meteorologie und Klimatologie nach den neuesten Forschungen*. A. Hartlebens, Wien (1891).

[59] Pierer Universallexikon: *Neuestes Wörterbuch aller Wissenschaften, Künste und Gewerbe*. Verlagsbuchhandlung von Ad. Spaarmann, Oberhausen (1877).

study[60] on the controlling influence of the natural environment on human behavior with the following programmatic generalization: "Man is a product of the earth's surface ... the earth has mothered him, fed him, set him tasks, directed his thoughts, confronted him with difficulties that have strengthened his body and sharpened his wits, given him his problems of navigation and irrigation, and at the same time whispered hints for their solution. ... Man can no more be scientifically studied apart from the ground he tills, or the land over which he travels, or the seas over which he trades, than polar bear or desert cactus can be understood apart from its habitat."

Another example is found in the influential work of the social psychologist Hellpach[61]: "In the Northern part of a region of the earth the traits of sobriety, astringency, coolness, serenity, the willingness to make an effort, patience, tenacity, rigor, the consistent use of the intellect and will predominate — in the Southern part the traits of liveliness, excitability, impulsiveness, the sphere of feelings and fantasies, the more sedate letting go or instantaneous flaring up. Within a nation, its northern populations are more practical, more reliable, but more inaccessible, its southern populations more musical, more accessible (cozier, more amiable, more talkative), but more volatile." Such ideas live on even today.[62]

Despite their increasingly quantitative orientation, climate determinists favored a kind of essential knowledge of climate. One trusts and invokes theoretical premises in which certain situationally independent or supra-personal properties (or traits) are primarily attributed to climate. These traits are held responsible for the fact that climate holds a comprehensive position of power over situation-specific, historical processes in almost every historical and social context. In the context of such an understanding of the natural climate, the "logic" of the respective social situation is at best of marginal importance for the explanation of its particularities and also its lines of development. Climate determinism, like other essence perspectives (race, technology, masculinity), loses its primacy after World War II. The social sciences, insofar as climate determinism still had any

[60] E. Semple: *Influences of Geographic Environment, on the Basis of Ratze's System of Anthropogeography*. Holt, Rinehart and Winston, New York (1911), especially pp. 1–2.

[61] W. Hellpach: "Kultur und Klima," in H. Wolterek (Hrsg.): *Klima - Wetter - Mensch*. Quelle & Meyer, Leipzig (1938), pp. 417–438, especially pp. 429–430.

[62] J. Pennebaker, B. Rime, V. Blankenship: "Stereotypes of emotional expressiveness of northerners and southerners: A cross-cultural test of Montesquieu's hypothesis," *Journal of Personality and Social Psychology 70 (1996)*: 372–380.

significance here at all, but also the religious sciences[63] and geography[64] increasingly emancipated themselves from essence perspectives. Nevertheless, there are again and again cautious approaches in the scientific literature,[65] which make clear that the concept has been suppressed but not destroyed and that one is interested in a rehabilitation of climate determinism.

2.3 Ellsworth Huntington

> *The people of Salzburg have always been as awful as their climate, and when I come to this city today, not only is my judgment confirmed, but everything is even more awful. … The climate of the foothills of the Alps makes people sick of mind, who fall prey to stupor at a very early age and who become mean spirited with time. … This climate and these walls kill sensitivity. …*
>
> *Thomas Bernhard*

Ellsworth Huntington (1876–1947) was born in Galesburg, Illinois, on September 16, 1876, the son of a Congregational minister. He attended high school in Maine and Massachusetts. As a young man, but also later in life, he traveled extensively in the Middle East, Asia, Europe, Africa, and North and South America. Huntington became a student at Beloit College in Wisconsin; he earned his doctorate in geomorphology or "physiography" at Harvard University. In 1907, he became an instructor in geography at Yale. There he was awarded the degree of doctor of philosophy in 1909, and in 1910 he was appointed assistant professor. In 1915, Huntington left Yale, returning in 1919. He did not teach geography, however, but spent 28 years as a Research Associate.[66]

Huntington was the best-known American geographer of the first half of the 20th century. For example, Social Education (Vol. XXI No. 1 January 1957) states: "Huntington was the American Geographer who was most widely known among educated people throughout the world. He studied with especial intensity weather and climate, their influences and changes, and greatly increased public interest. By many, Huntington was rated in his later years as the world's greatest geographer.

[63] K. Hoheisel: "Religionsgeographie und Religionsgeschichte," in H. Zimmer (Hrsg.): *Religionswissenschaft. Eine Einführung*. Dietrich Reimer Verlag, Berlin (1988), pp. 114–130.
[64] C. Troll: "Die geographische Wissenschaft in Deutschland in den Jahren 1933 bis 1945. Eine Kritik und Rechtfertigung," *Erdkunde* 1 (1947): 3–48.
[65] P.B. Sears: "Climate and civilisations," in: H. Shapley (ed.), *Climatic Change. Evidence, Causes and Effects*. Harvard University Press, Cambridge (Mass) (1953), pp. 35–50.
[66] G. Martin: *Ellsworth Huntington. His Life and Thought*. The Shoe String Press, Hamden, Connecticut (1973).

He certainly aroused more interest in geography on the part of more people than any other geographer."

In his investigations of the interaction of climate and society, Huntington was by no means satisfied with purely descriptive documentation or explanatory approaches. He was always eager to make practical recommendations, and he advocated in public that his conclusions be put into practice. For example, shortly after the Second World War, he recommended that the United Nations headquarters be built in Newport, Rhode Island, because the climate there was best suited to human needs.

In addition to geographical textbooks and his writings on eugenics, Huntington focused on the causes of the development or backwardness of human civilizations. This goal of knowledge took its first shape in 1914/1915 and found expression in a large number of scientific publications, the basic idea of which changed little over the next three decades.

Huntington's main thesis[67] about the effects of climate are as follows: "Climatic conditions constitute a distinct optimum (and conversely a downside) and with it varies the advance of civilization and the quality of people". Accordingly, the development of civilizations as well as the characteristics of people cannot be understood separately and independently of climatic conditions. A civilizational development favored by climatic conditions is evident in "[our] increasing ability to dominate the forces of nature. … Is it mere coincidence [he asks] that the English can fly in the air, sail beneath the ocean, manufacture machines by the million, and talk by radio, while not a man among the Kamchadales ever thinks of doing these things?" Progress he thus sees as a gift of nature.

Today, Huntington's notion that "climate paints the fundamental colors on the human canvas" may be seen as amusing, while others may consider it exaggerated, irrelevant, or absurd, and place it, at best, at the extreme edge of the social scientific discourse on the effect of environmental factors on human living conditions. At the time, however, Huntington's explanations were by no means atypical and could easily be fitted into commonly held social ideas about climate, health, and ethnic or racial identities.

Indeed, Huntington's evidence is essentially that he appeals to these widespread traditional assumptions and everyday prejudices: The variations in people's strength from month to month are so important and teach us so much about the distribution of health and energy throughout the world that we will study them closely … let us consider how physical strength varies during the course

[67] E. Huntington: *Mainsprings of Civilization*. John Wiley & Sons, New York (1945), especially p. 313.

of the year in the great section extending from southern New England and New York westward to the Rocky Mountains. October is usually the best month. At that time people feel like working hard; they get up in the morning full of energy, and go at their work quickly and without hesitation; they walk briskly to business or work; and play with equal vigor. Headaches, colds, indigestion and other minor illnesses are fewer than at other seasons; there are also fewer serious illnesses, so that doctors have less than usual to do, and the number of deaths is less than at any other time of the year." With the exception, perhaps, of the last statement, these are observations that seem to recall everyday experience and are considered easy to understand and follow. The same applies to the conclusion that there is a "well-known contrast between the energetic people of the temperate zone and the lazy inhabitants of the tropics. It is taken for granted and proven that "*everyone is influenced by temperature, humidity, wind, sunshine, barometric pressure, and perhaps other factors such as atmospheric electricity and the amount of ozone in the air. On days when all these factors are favorable, people feel strong and hopeful; their bodies are capable of unusual exertion, and their minds are alert and accurate. If all the factors are unfavorable; people feel inefficient and dull; their physical weaknesses are exaggerated; it is hard to concentrate the mind; the day's work drags slowly; and people go to bed at night with a tired feeling of not having accomplished much. Hence in a variable climate like that of the United States people's physical and mental energy keep changing from day to day and season to season. Sometimes one feels almost as inert as if he lived within the tropics, but soon a change comes and once again feels the health and energy which makes it possible to work hard and think clearly*."

It may well be that people feel differently in the course of the year; but whether this is connected with the temperature or the length of the day or in particular with social characteristics such as exam times, harvest time, family festivities, remains to be seen. It may also be that sunshine has an inspiring effect after the passage of a cold front.

However, Huntingtonian climate determinism develops its full persuasive power only through its pronounced generalization. For example, after Huntington has empirically established that the highest productivity is observed in factories in the New England states of the USA when outside temperatures are around 18°C, he concludes that 18°C is an optimum independent of location, social and cultural order.[68] Provided that the climatic conditions fluctuate in the course of the year in a not too wide band around this optimum and the weather conditions are not too monotonous,

[68] E. Huntington, W. Sumner: *Principles of Human Geography*. John Wiley & Sons, New York (1921), in particular p. 249.

one speaks of optimal climatic energy. This climatic energy permits the civilizational development, whereby, completely in the sense of the being-like interpretation of the power of the climate, no historically special characteristics, as, for example, those of the social organization, the economic system or the cultural coinage for the development of civilizations have an important meaning.

It is a characteristic of quantitatively oriented climate determinism that correlations between health or economic measures and the annual cycle are calculated. If one disregards the fact that correlations of this kind, as Durkheim already noted with certainty, do not have to describe causal relationships anyway, the application to deterministic cycles is statistically inadmissible. If one follows this logic, one could also claim that the peasants of North Germany are happy in autumn because the storks have left the country for Africa. It is equally wrong to infer from the effects of short-term weather fluctuations in one place the effect of other climates on other people in another place. A cold snap of a few days does not allow Hamburgers to experience a Greenlandic climate. Both arguments are often used by Huntingtonian climate determinists to generalize everyday experiences and are accepted as plausible by laymen.

Interestingly, not only the generalizations but also the locally established correlations are inapplicable. The fluctuations from year to year, for example, in piecework are larger than the fluctuations within a year — and do not correlate with temperature variations. The claim that economic productivity is related to weather fluctuations is rejected by economists.

The generalizations — and this includes Huntington's merely rhetorical reference to "other" factors such as cultural processes — are of such a general nature and must be subject to so many caveats that they preclude a sustainable qualification. Thus, when necessary, reference is made to other or "also playing a role" forces, delimitations, or exceptions, such as: "Thus, if all other influences were eliminated, we should expect civilization to advance most rapidly in climates which have few or no months with temperatures above the optimum and many below, but none too far below the optimum. As a matter of fact, the actual distribution of civilization approaches this pattern but departs from it in some respect because mean temperature is only one of the climatic factors of environment, and the effects of physical environment are modified by cultural environment."[69]

In this quotation, Huntington refers to one of his key arguments, namely, that his geographic determination of "climatically favorable" or "unfavorable" regions corresponds to a global distribution of a "high" or "lower" level of civilizational development derived by expert opinion. Seen entirely through European eyes, a

[69] E. Huntington: *Mainspring of Civilization*. John Wiley & Sons, New York (1945), p. 275.

high level of civilization is only attested where Europeans settle, that is, essentially in Europe itself, North America, and Australia. The possibility that this coincidence could have other than climatic causes is not considered. Huntington omits to elaborate the effect of other non-climatic factors, so that climate stands as the only truly *independent* variable.

Huntington also addressed the issue of slow climate change in historical times. At that time, climate changes were essentially understood as either anthropogenic, through changes in land use, or cyclical. Huntington saw these changes as causes for the spatial displacement of advanced civilizations, especially in Asia: "*cyclic [climate] variation seems to have influenced some of the greatest historical changes. The highest mental achievement is possible only when favorable conditions exert a combined stimulus. Our task just now is to separate climatic effects from those of heredity, culture and the non-climatic physical environment.*"[70] The rise and fall of civilizations were seen as closely related to climate: "*In the regions occupied by the ancient empires of Eurasia and north Africa, unfavorable changes of climate have been the cause of depopulation, war, migration, the overthrow of dynasties, and the decay of civilization; while favorable changes have made it possible for nations to expand, grow strong, and develop the arts and sciences.*"[71]

The Middle Ages and the Renaissance in Europe are interpreted as an example of slow changes in climate and the associated idea of climate impact: "*The Dark Ages and the Revival of Learning occurred at opposite phases of a long climatic cycle. Storminess apparently reaches a long ebb in the Dark Ages but an abundance and violence in the fourteenth century. These two periods were likewise times of psychological contrast. The Dark Ages were characterized by widespread depression of mental activity, whereas the Revival of Learning ushered in a period of alertness and hope.*" [72,73]

Huntington also raised the possibility of global climate change, which he considered rather unlikely: Even if one disregards a possible dramatic decline of developed civilizations into a state of backwardness, such as tropical societies show according to Huntington, the future prospects of mankind are bleak, since changes in the geographical position of the world's population are not likely to lead to a global climate change since the "*highest climatic energy and the consequent rise of new powers and the decline of those now dominant may throw the*

[70] E. Huntington: *Mainspring of Civilization*. John Wiley & Sons, New York (1945), p. 344.

[71] E. Huntington: *Palestine and Its Transformations*. Houghton, Mifflin & Co., New York (1911).

[72] E. Huntington: *Mainspring of Civilization*. John Wiley & Sons, New York (1945), p. 343.

[73] *Ibid*, p. 345.

world into a chaos far worse than that of the Dark Ages. Races of low mental caliber may be stimulated to most pernicious activity, while those of high capacity may not have energy to withstand their more barbarous neighbors."

Huntington's ideas about climate research are consistent with the efforts of the pioneers of the newly emerging scientific disciplines of meteorology and climatology. The latter were concerned with finding objective measures to describe the climate. Huntington consistently applied this idea to climate impacts. He considers himself more modern and scientific than his predecessors Montesquieu or Falconer, whom he does not mention because he focuses on "hard," quality-assured and reality-conforming climate statistics. While his predecessors had to rely on rather vague subjective perceptions and random reports, Huntington works with instrumental data and could therefore claim a scientific standard for himself and his theses that corresponds to the times.

From the point of view of modern social science, the Huntingtonian climate determinism implies assumptions that contradict certain widely accepted essential premises about social behavior among social scientists. Among these assumptions of climate determinism, we include the stability of climate and social behavior and the alleged "egalitarianism" of climate.

The stability of climate and social behavior contradicts the observation that (modern) social reality has a rather fragile, dynamic character and is in constant change. Since, in Huntington's view, climate is constant except for slow changes, the consequences of climate are also static and extremely stable life-worlds emerge.

Nowadays, most social phenomena are usually understood as partly stratified processes. However, climatic energy affects everyone equally, regardless of social status, wealth, or political influence. The advantages and disadvantages of Huntington's climatic energy affect all people in a climatic region equally, regardless of the social and cultural factors that social scientists usually refer to when discussing the causes of social change, individual identity, social mobility, or social inequality. Huntington's classic climate determinism reverses the causality of the relationship between climate and society and thus underestimates not only societies' chances of emancipation from climatic circumstances, but also the socially differentiated indirect influence of environmental conditions.

2.4 Eduard Brückner

Eduard Brückner was born in Jena in 1863 and died in Vienna in 1927. He studied geography, geology, paleontology, physics, meteorology and history in Dorpat,

Dresden and Munich. In 1885 he completed his doctorate under the guidance of Albrecht Penck with the topic "The glaciation of the Salzach region." This was followed by a brief stint at the Seewarte in Hamburg before he became professor of geography at the University of Bern in 1888, of which he was rector in 1899/1900. After a short stay in Halle, he became Penck's successor in Vienna and represented climatology there together with Julius von Hann. Brückner achieved fame both for his work on ice ages in the Alps and for his hypothesis of 35-year quasi-periods, which were also called Brückner periods. In the *Encyclopedia Britannica* at the end of the 1920s, Brückner's name is mentioned under the heading "Climate" as one of a few climatologists.

Brückner's main work "Klimaschwankungen seit 1700 nebst Bemerkungen über die Klimaschwankungen der Diluvialzeit" was published in 1890.[74] In this study, he first dealt with the frequent contradictions in the literature, according to which there were countless reports of an increase in precipitation, but just as many of a decrease in precipitation, as well as the numerous hypotheses as to the extent to which these changes were anthropogenic (in particular through a change in land use) or "cyclical" (i.e., of natural origin) in nature. He then worked out in detail and data-critically that the contradictory reports were an expression of natural fluctuations with a quasi-period of 35 years, and reconstructed climate fluctuations up to 1700 and further back. He rejects the hypothesis that human activity is changing the climate. He operated not only with instrumental data but also with proxy data such as wine harvests and the like.

In the present context, however, Brückner is particularly interesting because of his studies on the social consequences of natural climate fluctuations. He examines the effects of transport on rivers (icing), on health (e.g., typhoid fever) and on agricultural production.[75] Brückner states a clear influence, which, in turn, is strong enough to affect migratory movements from Europe to North America[76] and the global power relations of the continental and maritime powers of Europe. He assumed that the relationships once found would remain valid in the future and were therefore suitable for forecasts that should be valid in principle for the periods of about 35 years that he identified. Statements about changes in the world economy and in the world political balance of power should be possible.

[74] E. Brückner: *Klimaschwankungen seit 1700 nebst Bemerkungen über die Klimaschwankungen der Diluvialzeit*. E.D. Hölzel, Wien und Olmütz (1890).

[75] E. Brückner: „Der Einfluß der Klimaschwankungen auf die Ernteerträge und Getreidepreise in Europa," *Geographische Zeitschrift* 1 (1895): 39–51.

[76] E. Brückner: "The Settlement of the United States as Controlled by Climate and Climate Oscillations," Memorial Volume Transcontinental Excursion of 1912 of the American Geographical Society of New York (1915).

Here Brückner resembles modern climatologists, who do not assume natural, transient, but stress anthropogenic, *permanent* climate changes, if he was convinced that once observed climatic effects will also be observable in future societies future. In fact, this was not the case. The growing dominance of railroads marginalized the economic importance of river navigation, and disruptions caused by river freezes became largely insignificant. The development of medicine, especially hygiene, made supposedly climate-dependent diseases and epidemics manageable. The success of breeding and the possibilities of artificial irrigation made agriculture less dependent on climatic fluctuations. The indirect effects, if they can be reliably specified, also became increasingly insignificant: People no longer emigrated from Europe to the USA because of the economic situation, but because of the political situation, and globally relevant political conflicts of interest no longer pitted continental and maritime European powers against each other.

In short, the effect of climatic conditions on economic and social processes and changes that Brückner researched and predicted did not occur because social and technological conditions were rapidly changing. And as the momentum of social, scientific, and technological change continues to increase,[77] the importance of the "non-natural" as a driver of social change will diminish and human emancipation from climate will increase. However, this does not at all mean that the influence of the social construct, that is, the widespread social understanding of climate and climate change on politics, science and society is marginal.[78]

2.5 The self-limitation of social science discourse

As the introductory example of Durkheim illustrated, all those theoretical perspectives that directly refer to the influence of natural forces as explanatory variables for social processes were successfully excluded from the social science discourse. And for this exclusion, it should be emphasized, there were good reasons then as now. The main ones are the following: (1) biological and cultural development are not identical; (2) the natural environment of society is to a large extent independent of human action, and societies have been successful in attempting to free themselves from many of the constraints of the environment.[79] As a result of this

[77] N. Stehr: *Die Zerbrechlichkeit moderner Gesellschaften: Die Stagnation der Macht und die Chancen des Individuums in der Wissensgesellschaft.* Velbrück Wissenschaft, Weilerswist (2000).

[78] N. Stehr, H. von Storch: "The social construct of climate and climate change," *Climate Research* 5 (1995) (see Chapter 2 in this anthology).

[79] R. Grundmann, N. Stehr: „Klima und Gesellschaft, Soziologische Klassiker und Außenseiter. Über Weber, Durkheim, Simmel und Sombart," *Soziale Welt* 47 (1997): 85–100.

differentiation, social science discourse then also largely succeeded in resisting the ever-elusive simplicity of most theoretical explanatory models of technological, economic, and biological determinism.

The history of the social sciences in this century can therefore be understood as a struggle against social Darwinism, racism, climate determinism, and against sociobiology. The dominant social science disciplines succeeded in limiting their discourse to *sui generis* processes, be they social, political, economic or cultural. The question of the necessary social conditions for social order became a fundamental topic for social theorists. The ecological conditions for social order were considered unproblematic or were assigned to other academic sciences in the sense of an intellectual division of labor.

The social sciences not only consciously discarded all references to physical, biological, and other environmental factors because they were anxious to institutionalize their own independent perspectives and problem areas, which were unequivocally different from those of the natural sciences. They also shared, to a large extent, certain ideological or moral premises that were closely linked to the idea of modernity and development. These included, in particular, the conviction that the transformation to modern society and desirable living conditions involved extensive emancipation from the immediate influence and dependence on environmental factors. The liberation from (reductionist) naturalism is therefore a kind of intellectual emancipation.

At present, however, society's influence on nature and, much less urgently, nature's influence on society is at the forefront of many discussions in science and politics. Moreover, nature is also described in scientific discourse as *changeable* and its sensitivity to human intervention is examined. The emancipation of society from nature is followed by the paradoxical development that this emancipation creates a new dependence, as in the case of anthropogenic climate change.

This observation could not be expected directly from the social sciences, but the study of these new dependencies cannot remain the sole object of the natural sciences. The social sciences are therefore forced to reconsider their relationship to nature unless they can be completely excluded from these emergent fields of research. Thus, the social sciences are faced with the task of renewing and transforming the concept of nature in social science discourse.

Here, on the one hand, the errors of any (reductionist) naturalistic determinism, such as climate determinism, must be avoided.

The mere introduction of the problem field "environment" into the discourse of the social sciences, as if it were nothing else than a topic like divorce or unemployment, is insufficient. So far, it has not been possible to satisfactorily redefine the paradigmatic relationship between society and nature in social science discourse.[80]

Approaches to bringing nature and society together in social science discourse include Bruno Latour's program[81] to overcome the dualism of nature and society, the various works of feminist eco-sociology, or those of neo-Marxist theorists.[82] We believe, however, that this can only be achieved in a sustainable way if the pull of naturalism or those concepts based on a purely constructivist perspective is resisted and the traditional scientific division of labor is overcome in order to create a kind of *social natural science*[83] in a transdisciplinary way, in which both natural conditions and their changes and our observations of them are understood as social processes for society, nature, and research.

2.6 Outlook

After we have tried to present the spectrum of ideas of climate determinism on the basis of Huntington's and Brückner's theses, we want to turn to the question posed in the title: Is climate determinism only a history of ideas or a relevant factor in current climate policy? In our opinion, climate determinism is, for a number of reasons, a relevant development in the history of ideas today, which should interest both the social and the natural sciences.

The basic assertion that climate and climate change directly determine and control social conditions and developments in modern societies cannot be sustained. The assertion of the direct influence of climatic factors compared to social processes independent of the natural environment is too strong. The geographer Wilhelm Lauer put it this way: "The climate is indeed of importance for the shaping of the scene on which human existence — the history of mankind — takes place, because it defines the framework in the broadest sense, limits the possibilities, sets boundaries for what can happen on earth, but not what happens or will happen. At best, the climate poses problems that man has to solve. Whether he solves them, and how he solves them, is left to his imagination, his will, his creative activity. Or expressed in a metaphor: the climate does not write the text for the development drama of mankind, he does not write the script of the film, man

[80] W. van den Daele: "Concepts of nature in modern societies and nature as a theme in sociology," in: M. Dierkes, B. Biervert (eds.), *European Social Science in Transition. Assessment and Outlook.* Campus, Frankfurt am Main (1992), pp. 526–560.

[81] B. Latour: *We Have Never Been Modern*. Harvard University Press, Cambridge (Mass) (1993).

[82] For example: A. Gorz: *Capitalism, Socialism, Ecology*. Verso, London (1991, 1994).

[83] N. Stehr, H. von Storch: "Soziale Naturwissenschaft oder: Die Zukunft der Wissenschaftskulturen," *Vorgänge* 37 (1998): 8–12.

alone does that."[84] Climate defines a framework or conditions of action within which social dynamics can develop. In the course of historical development this framework widens.

The thesis of the direct impact of climate on society can also be called the "naturalization" of climate impacts. A recent example can illustrate this:[85] In mid-July 1995, the American megacity of Chicago experienced one of its greatest heat waves and, in its wake, the deadliest environmental disaster in the city's recent history. For nearly a week, an unusual but predicted large-scale weather pattern was responsible, providing temperatures of 41°C and a heat index (i.e., a combination of temperature and humidity) of 49°C. Skies were clear, there was no cooling wind from Lake Michigan, and even the daytime lows were dangerously high. The heat wave caught the city off guard. During the week-long heat wave, 739 more people died in Chicago than the average for that week in July in the past. A forensic investigation concluded that more than 500 people were direct victims of the extreme temperatures. Are the dead victims of an extreme weather event in the city of Chicago, is Chicago soon to be everywhere, or is the city's social order responsible?

In reality, new forms of social marginality were responsible, such as the increased isolation of older people or the concentrated poverty of certain Chicago residents, who were much more out of shape in the mid-1990s than they had been in the middle of the previous decade. These are socially influenced and influenceable structures of risk and vulnerability that are crucial to such effects. It is the social construct of vulnerability that transforms natural weather extremes — for certain individuals — into disasters. The major city of Milwaukee, 150 kilometers from Chicago, counted 91 (heat) deaths in the same week of July 1995, using the same yardstick. But even in Milwaukee, the extreme temperature is by no means causal or even solely responsible. However, naturalizing the consequences of the heat wave may well be in the interest of the political class. These findings and the inadequacies of classical climate determinism have direct consequences for modern climate impact research. A "denaturalization" of climate impact research is necessary. Moreover, purely scientific scenarios cannot provide meaningful estimates for the future with regard to society and the economy, since the most important factor, the dynamic development of society and technology, is not taken into

[84] W. Lauer: "Klimawandel und Menschheitsgeschichte auf dem mexikanischen Hochland," Akademie der Wissenschaften und Literatur Mainz. Abhandlungen der mathematisch-naturwissenschaftlichen Klasse 2 (1981).

[85] E. Klinenberg: "Denaturalizing disaster: A social autopsy of the 1995 Chicago heat wave," *Theory and Society* 28 (1999): 239–295.

account. But it is precisely the estimation of future societal development lines that is associated with great uncertainty.

In our opinion, it follows from these considerations that adaptation strategies and research are practically more effective than mitigation research and strategies.

Acknowledgments

We thank Robert Antonio, Kevin Haggerty, Gerd Schroeter, Volker Meja, and Jay Weinstein for their constructive criticism of the first version of this study. We also thank Sönke Rau for his rough translation of the original English version of the text into German and Barbara Stehr for her editorial revision. We dedicate this paper to our colleague Gerd Schroeter of Lakehead University in Thunder Bay, Ontario, Canada, who died suddenly in the spring of 1999. We lose in him a scholar who, over many years, displayed a unique collegial willingness to read the emerging texts of others with great care. His ability to critically work through manuscripts with constructive thoroughness and thus bring them to life was unsurpassed. This essay, like others in the past, has benefited from Gerd Schroeter's generosity. We are grateful to him.

3. Eduard Brückner's Ideas — Relevant in His Time and Today[‡]

3.1 Introduction

3.1.1 *Temporal flow of ideas and the failure of diffusion*

For a natural scientist, scientific discourse develops like the trunk of a tree.[86] Each year, a new tree ring is formed based on the most recent findings incorporating previous results — from the most recent tree ring, so to speak — and newly established facts and interpretations. Knowledge obtained from earlier research is either encoded or obliterated in present knowledge continuously transferred from tree ring to tree ring — or forgotten. If something has not been incorporated from cohort to cohort of scientists, it is considered to be irrelevant and of little interest. This approach in natural science to its own history is clearly manifest in practically all contemporary articles in scientific journals. Most of the citations refer to work not older than 5 years. Sometimes casual reference is made to a handful of "classical" papers or books, but the authors have likely never closely examined these classic works but know about them only indirectly.

This mode of operation is undoubtedly an efficient way of coping with the sheer number of publications scientists face daily. It is simply not possible to digest all new results — even in a field as relatively narrow as climate science — let alone critically read many of the potentially relevant original documents of past research. For example, for the process of understanding a map displaying the global temperature by means of isotherms, it is not important to know that the

[‡]This section originally appeared in Stehr, N. and H. von Storch: "Eduard Brückner's ideas — relevant in his time and today," in N. Stehr and H. von Storch (eds.), *Eduard Brückner — The Sources and Consequences of Climate Change and Climate Variability in Historical Times*. Kluwer, 2000.

[86]We use concepts geared towards the thinking of physicists. The relevant background is the transport of heat in a fluid. Heat can be transported either by *diffusion*, which is maintained by the collisions of the individual molecules within the fluid. This transport is relatively inefficient as any transport is made up of many little steps over the small distances between two molecules. In our metaphor, this means that knowledge is transferred through personal contacts among scientists and through recent publications. A more efficient way takes place when a current transports a parcel consisting of many molecules as a whole over a longer distance. This process is called convective transport. In our context, this refers to the introduction of forgotten concepts and results into contemporary thinking.

technique of isotherms was invented by Alexander von Humboldt, or what his ideas about the technique were at the time.

In almost all cases, this "diffusive" transfer of knowledge from cohort to cohort and generation to generation, or from "tree ring" to "tree ring," works effectively and is robust enough to filter out what are considered to be irrelevant constructions from the flow of knowledge. At the same time, during this process of consensus building all knowledge claims are continually and critically examined with respect to recent insights. Today, no one in the natural science community would advance claims based on old authoritative sources, as was common, for example, with the work of Aristotle during medieval times. However, this process is not effective when a line of inquiry in science is displaced for some reasons — and interest then reappears after a longer period of time has passed and the collective memory about past intellectual perspectives is no longer available in present-day journals and scientists. In such a case, the transfer of knowledge needs more than "diffusion" but outright "convective transport" from deeper placed tree rings to the surface.

This "convective" transfer of ideas from the past should not be misunderstood as an attempt of merely repeating and preserving cognitive traditions. Hand-in-hand with the transmission of ideas from the past goes a mediation and interpretation of such ideas in the light of new circumstances and therefore present problems and issues. Thus, familiarity with past ideas can be instrumental in the construction of new knowledge and is not so much an obstacle to scientific discovery, as the practices of the scientific community today often appear to imply but an intellectual asset in efforts to advance science.

We believe that climate science is a case for which the "convective" influx of past ideas is a compelling necessity. After having been the bookkeeper for geography and meteorology in the 19th century, climatology developed into a science of the physics and chemistry of the atmosphere and the ocean; the early view that climatology is foremost a field of study that deals with the impact of climate on people and society was virtually forgotten. In the 1980s and 1990s, climate science underwent another paradigmatic change: after the discovery that humankind is about to change climate, the old problem of anthropogenic climatic change and the influence of climate on individual and society re-emerged.

During our own work on the interrelations between climate and social conduct we came across a number of early climate scientists who had a significant impact on both their peers and the general public. One of them was the eminent geographer Eduard Brückner, who is today forgotten in climate science, and is considered by geographers to represent but a closed episode in their disciplinary history.[87] At the turn of the twentieth century, he was one of the central protagonists in a vigorous debate in science and society about global climate variability and its political and economic significance. We believe that his formidable ideas could have a significant impact on our present view of climate, climate variability and climate impact. It is for this reason that we have assembled this anthology of Brückner's main work on climate variability and climate impact.

3.2 Organization of the section

In this section,[88] we present information about Eduard Brückner and his scientific work, compare his approach with that of his contemporary Julius von Hann, and relate his views to the present-day discussion.

The main part of this book consists of reprints of Brückner's original work in climate science. As most of his publications were in German, they were translated. These translations of Brückner's texts conform strictly to the original. Only in the case of completely irrelevant notes have we decided to delete such references. Additions we have made are inserted in square brackets. All diagrams have been redrawn. Some native city names used may be less familiar than the English names: München is Munich; Wien, Vienna; Praha, Prague.

The following is a list of the material presented. These eleven items were chosen as they demonstrate well Brückner's interest in climate variability, his assessment of contemporary analyses and thinking about anthropogenic climate change (such as the widespread concern about desiccation), and how he has dealt with the transfer of knowledge into society.

1. *Groundwater and Typhus [Grundwasser und Typhus. Mitteilungen der Geographischen Gesellschaft in Hamburg]*, Volume III, 1887–1888.
2. *Fluctuations of Water Levels in the Caspian Sea, the Black Sea, and the Baltic Sea Relative to Weather [Die Schwankungen des Wasserstandes im Kaspischen Meer, dem Schwarzen Meer und der Ostsee in ihrer Beziehung zur Witterung].* Annalen der Hydrographie und Maritimen Meteorologie, Volume II, 1888.
3. *How Constant is Today's Climate? [In wie weit ist das heutige Klima konstant?].* Verhandlungen des VIII Deutschen Geographentages, 1889.

[87] An informative overview about Brückner's scientific career and achievements can be found in Grosjean (1991).

[88] This section incorporates some materials first published by Stehr *et al.* (1995).

4. *Climate Change Since 1700 [Klimaschwankungen seit 1700.* Excerpts from *Klimaschwankungen seit 1700].* Wien: E.D. Holzel, 1890. Chapter 1: *The Current Status of the Inquiry into Climate Changes [Der gegenwärtige Stand der Frage nach den Klimaänderungen].* Chapter 8: *Periodicity of Climatic Variations derived from observations of ice conditions on rivers, the date of grape harvest and the frequency of severe winters [Die Periodizität der Klimaschwankungen, abgeleitet auf Grund der Beobachtungen über die Eisverältnisse der Flüsse].*
5. *About the Influence of Snow Cover on the Climate of the Alps [Über den Einfluß der Schneedecke auf das Klima der Alpen].* Zeitschrift des Deutschen und Österreichischen Alpenvereins, 1893.
6. *Influence of Climate Variability on Harvest and Grain Prices in Europe [Der Einfluß der Klimaschwankungen auf die Ernteerträge und Getreidepreise in Europa].* Geographische Zeitschrift, 1895.
7. *Weather Prophets [Wetterpropheten].* Jahresbericht der Berner Geographischen Gesellschaft, 1886.
8. *An Inquiry About the 35-Year-Period Climatic Variations [Zur Frage der 35jährigen Klimaschwankungen].* Petermann's Mitteilungen, 1902.
9. *About Climate Variability [Über Klimaschwankungen].* Mittheilungen der Deutschen Landwirtschaftsgesellschaft, 1909.
10. *Climate Variability and Mass Migration [Klimaschwankungen und Völkerwanderungen].* Talk at Kaiserliche Akademie der Wissenschaften, Wien, 1912.
11. *The Settlement of the United States as Controlled by Climate and Climatic Oscillations.* Memorial Volume of Transcontinental Excursion of 1912 of the American Geographical Society of New York, 1915.

3.3 The climate scientist Eduard Brückner

3.3.1 *The life of Eduard Brückner*

They say it is observed in the Low Countries, that every five and thirty years the same kind and suit of years and weathers comes about again; as great frosts, great wet, great drought, warm winters, summers with little heat, and the like, and they call it the prime; it is a thing I do rather mention, because, computing backwards, I have found some concurrence.

Francis Bacon

Eduard Brückner was born on July 29, 1863, in Jena, Germany.[89] He lived for a while in Odessa, Russia, before moving with his parents to Dorpat (now Tartu, Estonia), where he spent most of his childhood. In 1879 he was sent to school in Karlsruhe, Germany. After graduating from high school, he studied at the universities of Dorpat, Dresden, and München. He attended lectures and seminars in geography, geology, paleontology, physics, meteorology, and history. In 1885, he completed his doctorate under the supervision of Albrecht Penck in München with a dissertation on the *Glaciation of the Salzach area (Die Vergletscherung des Salzachgebietes)* in Austria. In 1886, he moved to the Office for Marine Weather *(Seewarte)* in Hamburg to work with Wladimir Koppen. The first two of our translated articles originate from this early period of his scientific career. They pertain to the possible link between groundwater levels and the incidence of typhus, and the relationship between sea water level variations and weather conditions. On the strength of his dissertation, Brückner was appointed professor of geography at the University of Bern in 1888. He stayed in Bern for 16 years and became Rector of University of Bern in 1899/1900. During his stay in Bern, he lectured on various aspects of geography but also regularly offered public lectures. In 1904 he accepted an offer from the University of Halle in Germany and, in 1906, finally moved, as the successor of his former teacher Albrecht Penck, to the University of Vienna. Brückner died in Vienna in 1927 at the age of 64. While in Vienna, he was, as in Bern, engaged in the transfer of academic knowledge to the general public. He was chairman of a series of Public University Lectures *(Volksthümliche Universitätskurse).*

In 1890, he published the first extensive book-length discussion of recent climate fluctuations, that is, of climatic fluctuations in "historical times." Brückner (1894:1) credits the head of the Bavarian meteorological services, C. Lang, with the discovery of decadal scale climate variability in a study of the climate of the Alps. We have selected Chapters 1, 8, and 9 of this monograph (item 4).

After 1890, Brückner published only a few smaller articles on the observational evidence of climate variability (Brückner, 1895, 1902; reprinted as items 6 and 8). He explains the small number of articles on the observational evidence as the result of a lack of new and appropriate meteorological data on the issue. In the present-day context of particular importance, though, are his articles in which he speculated about the geographical and socio-economic impact of climate change, that is, the social consequences arising from the climate fluctuations, such as emigration, immigration, and migration patterns (Brückner, 1912; [1912] 1915; items 10 and 11), or on harvests, the balance of trade of countries and shifts in the

[89] Cf. Grosjean (1991) and Oberhummer (1927).

Eduard Brückner and Albrecht Penck in the summer of 1893 on an excursion near Flims (Graubünden, Switzerland). Taken from Büdel (1977).

political predominance of nations (Brückner, 1894, 1895, 1909; the last two reprinted as items 6 and 9).

He was convinced that the issue of climate change and its impact was both of considerable scientific merit and that future climate changes are of great importance to the well-being of society as well as for the strategic and economic balance of political and economic powers. He, therefore, presented his conclusions about serious repercussions associated with climate change anticipated for the end of the past century in the form of oral presentations addressing the general public and especially affected segments of the public, such as farmers. As a result, Brückner presented his initial findings on climate change not only to a congress of professional geographers in Berlin in 1889 (our item 3) but also a year earlier in a public lecture entitled *Is our climate changing?* at the University of Dorpat that was duly noted in the local press (Brückner, 1888). Later Brückner (1894, 1909; the last one is reproduced as item 9) published newspaper articles about the general issue of climate change as well as about its specific economic and social consequences. His work on climate variability was discussed at length in the contemporary press (e.g., *Neue Freie Presse*, Vienna, February 11, 1891).

As a result of these activities and the response they generated, Brückner's work on climate variability found a considerable echo among the scientific community of climate researchers (e.g., DeCoumy Ward, [1908] 1918), sociologists (e.g., Sorokin, 1928:120–124), geographers (e.g., Huntington, 1915:172–173; [1915] 1924:25), historians (e.g., Le Roy Ladurie, [1971] 1988:217,220) and physicists (e.g., Arrhenius, 1903:570–571), but to some extent also among the public at large, as is exemplified by the fact that he was often mentioned as an influential climate scientist in various encyclopedias until the 1950s. Huntington (1915:172) elevates Eduard Brückner to *"one of the chief European authorities on climate"* and credits him for having initiated a kind of paradigm shift in climate research: *"Since the publication of Brückner's widely known book on 'Climatic Changes Since 1700' there has been a strong and growing tendency to treat climate as a dynamic instead of a static geographical force"* (Huntington, 1916:192).

3.3.2 *Eduard Brückner's analysis of climate variability*

In the following section, we summarise Brückner's attempt to synthesize the observational evidence for global-scale synchronous climate variability from his limited data and limited computing power. Most of this synthesis is described in his 1890 monograph.

Brückner (1889:2) indicates that he was first alerted to the possibility of climate change, aside from information about shrinking glaciers in the Alps, as the result of observations about changing water levels in the Baltic, the Caspian and the Black Sea (item 2). The changes in the water levels appeared to follow a specific pattern. The rhythm of the changes resembled changes in the glaciers of the Alps.

In his detailed discussion of "recent" climate fluctuations, Brückner (1890) justified his approach by referring to the studies of E. Richter, C. Lang and A. Swarowsky. Richter concluded that the causes for the secular variations of one specific glacier (*Obersulzbachgletscher* in Austria) are wet and dry periods lasting for several years in that particular region. Lang showed that this result is valid for the entire Alpine region. Swarowsky stated a striking correlation between the variation of the water level of the Neusiedler See, a lake without any outlet near the Austrian–Hungarian border, and the secular variations of the glaciers in the Alps, thereby demonstrating that lakes without an outlet are excellent indicators of secular climate variability.

In his 1890 monograph on climate variability, Brückner started his analysis with a careful investigation of the worldwide largest "lake" with no outlet, the Caspian Sea. Brückner drew the conclusion that Lang's results not only hold for the Alps but may be extended to the vast catchment area of the Caspian Sea (Brückner, 1890:86). He found that the climatic variation followed a characteristic 35-year pattern, with wet and cool conditions alternating with dry and warm conditions.

This inductive method of extending results from a smaller region to a larger one is, by the way, typical for Brückner's approach and consequently he searches in data available from several other lakes without an outlet all over the world for signals of secular variations. Brückner states that the mere existence of water variations in the lakes allows the presumption that secular climate fluctuations take place in the corresponding catchments (Brückner, 1890:115). In a further step, Brückner applies the concept of linking water levels of lakes to the rainfall in the corresponding regions also to lakes with an outlet (*Flußseen*) and even rivers thereby stating the existence of a more or less synchronous climate fluctuation over the entire land mass of the world (Brückner, 1890:132).

The record of instrumental observations available to Brückner reached back for about 100 years. In these data, he identified a rhythm of 35-year alternating wet/cool and dry/warm episodes. In order to trace these characteristic climate fluctuations further back, Brückner also studied the observed data of the ice conditions of the rivers, the grape harvest and the abundance of strong winters. According to his data, Brückner was able to identify 25 quasi-periodic cycles of about 35 years length during the last 1000 years (Brückner, 1890:286).

He emphasized the fact that his mode of variability was not strictly periodic but that the alternating wet and dry periods lasted about 35 years on average. This fact is insofar noteworthy as in Brückner's years, it was the fashion to decompose time series of all sorts into its Fourier components in an attempt to describe the time series as a sum of predictable components developed. Obviously, Brückner stayed away from this fashion, which later was shown to be based on a simplistic misunderstanding of the mathematics of statistical time series.[90]

He speculated that the dynamical mechanism behind his quasi-oscillation would be related to some unknown solar forcing mechanism (Brückner, 1890:240,242) but was aware that no observational evidence for such an oscillation exists. In this context Brückner denied any connection between secular climate fluctuations and variations of sunspot activity (Brückner, 1890:242).

Based on this 35-year period oscillation, Brückner prognosed a dry period at the turn of the century (Brückner, 1890:286,287) with severe negative consequences in crops for continental regions, like Northern America, Siberia and Australia. It is noteworthy that this predictive scheme would have enabled Brückner to predict the "dust bowl" in the central part of the United States, which actually took place during the Thirties of the previous century.[91]

---

[90]The fascination with the notion of periodic cycles as a description and an explanation for the rise and fall of geological phenomena, of plants and animals as well as social and economic processes, was still a vibrant enterprise in science during Brückner's career. Sir N. Shaw's *Manual of Meteorology* from the mid-1930s featured a list of several pages length of various periods found in meteorological data. The conviction that "the whole history of life is a record of cycles" (Huntington, 1945:453) was widespread. The fascination arises from the fact that a process made up of a superposition of a finite number of periodic sub-processes makes the process predictable: "It will be a vast boon to mankind when we learn to prophesy the precise dates when cycles of various kinds … will reach definite stages" (Huntington, 1945:458). In the 1920s and 30s, the Russian mathematician Slutsky showed that the Fourier analysis of a statistical time series always reveals some periodicities, even if the time series is constructed free of such periodicities. If different chunks of such time series are analyzed, different periodicities pop up and vanish. In spite of this finding, which today is completely understood, in some circles, and in particular among lay scientists, the interest did not cease. On the contrary, in 1941 the interest in the study of cycles led to the formation of a "Foundation for the Study of Cycles" by Edward R. Dewey. The Foundation exists to this day and claims to have more than 3000 members.

[91]In 1915, Brückner predicted that by 1920 "we may expect a maximum of humidity" in the United States (Brückner, 1915:132). This prediction exploited two pieces of information: First, the dynamical insight about the existence of a 35-year oscillation and second, Brückner's finding that precipitation was at its minimum around 1900. On the continental scale, his forecast was incorrect (Bradley, 1987: Fig. 6), but in a regional sense his forecasts were consistent with actual developments: The Great Salt Lake exhibited maximum lake levels from 1910 to 1930. Brückner did not spell out another prediction based on the same reasoning, namely that in the middle of the 1930s the

Brückner's methods were mainly limited to the exploratory statistical analysis of time series since confirmatory tools such as confidence intervals or hypothesis testing were not developed in combination with what might be called common sense. He was unfamiliar with dynamical arguments (for instance, concerning the geostrophic wind, which was well known among meteorologists of those days) and he was unaware of theories concerning the general circulation of the atmosphere (he failed to acknowledge the different dynamic character of the tropics as opposed to the extratropical westerly regime).

A fact that is impressive for modem climate researchers, who are used to being supported by computers and digital data files, is the amount of computational work done by Brückner. It seems that he did all the calculations himself. He computed 5-year totals, called *lustrum*, and checked their consistence by comparing records from neighboring stations. When data at neighboring stations at some time begin to diverge, he concludes that one of the two records is contaminated by some artificial effects, such as displacement of the instrument (such as a water level meter). He tried to correct for such *inhomogeneities*, and calculated correlations between his various time series to establish the degree of similarity between them. The sheer work of just collecting the data, checking their consistency, and calculating their statistics must have been enormous, and hardly imaginable for a modern scientist. His methodical approach is similar to what is done today when, for instance, compiling records of the global mean temperature. The difference is, of course, that the work is no longer done by human computers, but electronic hardware supervised by humans.

3.4 Climate change. Climate policies and society

> *The number of hypotheses and theories about climate change are numerous. Quite naturally they have caught the public attention, as any proof of past climatic change points to the possibility of future climate change, which inevitably will have significant implications for global economics.*
>
> *Eduard Brückner (1890:2)*

Today, the concepts of "climate variability," "climate change," and "climate impact" attract an enormous interest not only in the climatological,

United States would again suffer from dry conditions. Indeed, the Great Salt Lake exhibited a sharp water level drop in the early 1930s. Also, the "Dust Bowl" dry episode that led to persistent disastrous harvest failures in Central North America took place in the mid-1930s.

meteorological, and oceanographic community (von Storch and Hasselmann, 1996) but also in sciences concerned with climate-sensitive systems, such as biometeorology, ecology, coastal defense, or the social sciences. The discussion of "the climate problem"[92] is by no means limited to the scientific community. It has drawn a great deal of interest from a general public (Lacey and Longmann, 1993) perhaps haunted by anticipations of catastrophic developments as a consequence of future anthropogenic climate change (Stehr and von Storch, 1995). Evidence of a public and scientific preoccupation with "the climate problem" is given by such institutions as the "Intergovernmental Panel on Climate Change" (IPCC) and international conferences aimed at the establishment of International Climate Conventions.

The majority of the scientific and general public interprets the climate problem as a *new* challenge. Yet, although for much of the past two centuries most "climatologists" and meteorologists have been convinced, and have considered it to be almost an axiom, that global climate is a constant during historical times,[93] some 19th-century climatologists, geographers and meteorologists maintained that climate is *not* a steady phenomenon (e.g., Brückner, 1890; Hann, [1883] 1893:362), recognizing that climate varies not only on geological time scales (thousands of years and longer) but also on decadal and century time scales due to natural and anthropogenic processes.

The processes that were discussed at the turn of the last century as the source of climate variability and change were different. The "*natural* variability," unrelated to man's activities, was speculatively attributed to astronomical factors, such as the solar activity, and to processes in the interior of the earth. In addition, the

---

[92] We place the expression "climate problem" in quotation marks since it is not well defined. Natural scientists associate with this expression the understanding, prediction, and, possibly, control of climate variability. Social scientists, on the other hand, consider the perception of climate, and its social and political implications as the "climate problem."

[93] Brückner (1889:2) notes that during the 19th century, a distinct disciplinary division with respect to the issue of climate change could be observed: Geographers and geologists were more inclined to consider a persistent climate change to be a reality while meteorologists defended the thesis that climate is a constant. Brückner (1890:2) offers an explanation why most professional meteorologists and many geographers at the time were rather silent on the issue of climate change; as a matter of fact, he observes that they were embarrassed to engage in research and discussion about climate change. The reason for the reluctance is the wealth of competing hypotheses about climate change formulated earlier in the century. But previous efforts only resulted in many contradictory voices about the nature of climate change, so that climatologists then became reluctant to add to the cacophony of mere opinions. As late as 1959, the prominent climatologist H. Lamb complains that many of his contemporaries consider climate as something static (Lamb, 1959).

idea of deterministic periodic processes attracted much attention among climatologists. *Anthropogenic* "climate change" was thought to be the result of human activities, such as de- and reforestation or new cultivation of land in North America. The possibility that anthropogenic emissions of carbon dioxide might alter the global climate was first discussed by the chemist Svante Arrhenius (1896, 1903) but dismissed by him as a realistic perspective for the next few hundred years.

The intensive debate among climatologists at the turn of the century receded into the background when a new disciplinary consensus emerged that remained predominant until recently, namely that the global climate system contained overriding equilibrating processes providing resilience against secular climate fluctuations; fluctuations that did occur were seen as distributed around a fairly stable mean climatic condition. Any anomaly extending for a few years would be canceled by an opposite anomaly at another time. On average, nothing would change. One reason why the perception of climate variations on historical time scales became unpopular may be the rejection of "catastrophism" and the eventual acceptance of "uniformitarism" in geology, as proposed by Lyell in the 1830s. Some of the social scientific theories about the impact of climate on civilizations, for example, by Sombart ([1911] 1951:324; 1938), Ploetz (1911), or Hellpach (1938), are actually based on the explicit premise of constant climatic conditions (cf. Stehr, 1996). In the public arena at the time, other urgent issues and concerns displaced reflections about climate change and its impact on society.

In the following, we attempt to recover spirited discussions among geographers, meteorologists, and climatologists that occurred toward the end of the last and at the beginning of this century. We make an effort to analyze the dynamics of the discussion, and the degree to which it was introduced to the general public, with the explicit intention of comparing the situation at the time with the present discussions of climate variability and change and of climate policies designed to avoid or mitigate the risk of climate change or to allow for a smooth adaptation.

We concentrate on two of the main contributors to this early discussion of climate variability and change on time scales of decades, namely the already presented Eduard Brückner and Julius Hann, both professors in Vienna for a significant part of their lives. We will discuss their different social roles, their attitudes towards the role of the public, and their under- standing of their own work as part of multiple contexts in which they attempted to play different functions. We will show that the two protagonists, Brückner and Hann, represent roles and self-conceptions that resemble present-day roles of climatologists in discussions within and outside the scientific community about the scientific significance and the social impact of climate variability and change. We suggest that the "climate problem," as perceived by scientists and the public at the turn of the century, constitutes a valuable historical analog for present debates on the "climate problem."

3.4.1 *Julius Hann and his view of climate variability*

Another leading and most influential professional climatologist at the turn of the century was Brückner's Viennese colleague Julius Hann, who was born in 1839 in Wartberg, Austria. He studied mathematics, physics, geology, and geography at the University of Vienna. After a career in teaching, he became professor of physics at the University of Vienna and, in 1897, professor of meteorology at the University of Graz. Between 1900 and 1910, he occupied the newly created chair for cosmological physics at the University of Vienna and served as director of the Institute for Meteorology and Geodynamics. He died in 1921 at the age of 83 in Wien.

As Brückner (1923:152) points out in his obituary, Hann may well have been the most important meteorologist of his day and can be considered to be one of the founders of modern meteorology as the science of the physics of the atmosphere (see also Steinhauser, 1951; Kahlig, 1993). He was descriptively oriented, that is, keen to establish the empirical or observational basis for various meteorological phenomena. In meteorology, Hann discovered, independently of Helmholtz, the thermodynamic theory of the Föhn. In climatology and meteorology, he recognized early the importance of quantitative methods and the significance of three-dimensional observation systems, and he initiated the establishment of several mountain observatories. In addition, Hann was editor of the internationally recognized *Meteorologische Zeitschrift* for more than fifty years. Hann was an enemy of speculative thinking; his main goal was to establish the facts (Brückner, 1923:155).

Julius Hann compiled the first textbook on climatology. He first published his *Handbuch der Klimatologie* in 1883. The *Handbuch* appeared in a number of subsequent editions and translations and quickly became a classic in meteorology and climatology (cf. Brückner, 1923; Köppen, 1923:vi; Knoch, 1932:viii). An English edition based on the second edition of the German version of the *Handbook* was published in 1903 (Hann, 1903).

In contrast to later editions of the *Handbuch*, its first edition summarizing the state of knowledge in climatology, then still defined as an auxiliary science (*Hilfswissenschaft*) of geography (Hann, 1883:5; also Köppen, 1923:1), did not explicitly deal with the issue of climate variability. Reflecting the preoccupation of the day with the issue of periodicity of climate, Hann distinguishes between two types of climate fluctuations, namely "progressive" (i.e., persistent transformations, or, in modern terms "climate change"; e.g., von Storch and Hasselmann, 1996) and "cyclical" changes (i.e., fluctuations or oscillations around a constant mean with certain characteristic times or periods; in modern terms "climate variability"). The period for cyclical climate changes could be determined either by deductive reason (by postulating a certain forcing mechanism, such as the sun's activity) or inductive reasoning (by screening the observational record). It should be possible, according to Hann, to trace progressive climate change to either long-term trends of the temperature of the core of the earth or of the output of the sun.

As far as relevant empirical material is concerned, Hann ([1883] 1897:390). referred to both non-instrumental and instrumental observations of temperature and precipitation as well as general accounts or conclusions about climate changes of a wide variety of observers found in disparate historical records. He placed considerable emphasis on the critical examination of the observational climatic record. Obviously, such data can be used only if the procedure of observing, archiving and, possibly, correcting the raw data is kept constant (cf. Jones, 1995). The historical data available to Hann did in general not satisfy this homogeneity condition. He found on close examination that the data recorded in the previous 150 years were almost always contaminated by time-variable biases due to changing observational practices; the oldest instrumental records invariably were started in rapidly expanding cities, and therefore reflect "urbanization," while rain gauges were first placed on higher elevations (e.g., roofs) causing severe biases in measurements (cf. Karl *et al.*, 1993).

On the basis of such methodical pitfalls concerning the quality of the data, Hann was in general rather skeptical of scientific claims identifying climate variability and change in the observational record. In particular, he inferred that the evidence for systematic trends ("progressive changes") of the climate during the historical period based on the available data from different centuries, continents and countries is not substantial (e.g., Hann, [1883] 1897:390). It had been hypothesized that the continental United States of America of the 18th century was subject to an anthropogenic climate change due to the progressive anthropogenic transformation of nature in the course of colonialization. Hann concluded with Whitney (1894), that there is no hard evidence for a resulting climate change on the North American continent (Hann, [1883] 1897:392).

In the case of climate *variability*, Hann was less reluctant. He was skeptical about strictly periodic climate fluctuations, especially in regard to any hypothesized connection between variations in sunspot activities and meteorological elements such as temperature, precipitation or changes in the formation of glaciers. On the contrary, he concluded that the influence of sunspot activity on climate patterns is insignificant. Moreover, he rejected the possibility of any predetermination or causal linkage between climatic variations and sunspot activities (Hann, [1883] 1897:394).

Hann considered Brückner's quasi-oscillatory 35-year cycle much more favorably since it was based on rich data from very different sources. Brückner's discovery seemed valid for many regions and periods and was supported by many independent observations. Hann ([1883] 1897:400) made no serious independent attempts to clarify the dynamics of Brückner's observational evidence. Instead, he limited himself to efforts to establish the existence of the patterns of climatic fluctuations. Hann highlighted the fact that Brückner's observations manage to shed light on contradictory accounts of climate variations in specific localities since they "obviously" must have been advanced during different phases of the 35-year period.

Indeed, the second edition of the *Handbuch*, published in 1897, contains a forty-page separate section on climate variability that centers on Brückner's research. In the fourth edition of the *Handbuch*, published in 1932, Karl Knoch had succeeded Hann as author of the *Handbuch*, (Hann and Knoch, [1883] 1932). This fourth edition deals even more systematically with climate variability, even if the summary is rather skeptical. Much prominence is given to contributions that attempt to demonstrate the stability of the climate in historical times and point to the absence of evidence for secular change (see also Berg, 1914).

3.4.2 *Climate variability and societal importance*

It was and is common sense that climate variability and climate change have a direct and powerful effect on many aspects of society, including the economy, human health, or even the balance of power among nations.[94] Based on these views, the perspective of *Climate Determinism* emerged suggesting that climatic conditions *determine* virtually all aspects of social life, especially the chances of a society to attain a "high level of civilization."[95] This approach was widely

[94] The impact of climate on the course of history has been of considerable interest. A more recent account is Lamb's monograph *Climate, History and the Modern World* from 1982 (second edition 1995). Classical views have been put forward by ancient Greek authors such as Hippocrates and philosophers of the Enlightenment such as Montesquieu and Herder. Also, Friedrich Engels theorized about the influence of climate on society.

[95] The perhaps most prominent representative of modem climate determinism is Ellsworth Huntington (1915, 1945). For a discussion of the climatic determinism, see Stehr and von Storch (1998).

accepted in geography and other disciplines at the time of Brückner and Hann.[96] It is therefore of interest to inquire how Brückner and Hann responded to the challenge of offering their findings to the scientific and general public as warnings of impending climate change but also as instruments to design strategies to deal with climate variations. Interestingly, the two scientists reacted very differently.

Hann disregarded societal impact entirely. He did not even mention possible social consequences of climatic fluctuations. Consistent with the then-prevailing self-conception of climatology as a largely descriptive (e.g., Hann and Knoch, [1883] 1932:3) "young" science (e.g., Köppen, 1923:v), Hann examined the existing evidence on climate variability and change and attempted to establish whether the data supported arguments for changes in climate phenomena.

Brückner, on the other hand, not only discussed the nature and extent of climatic fluctuations but emphasized their possible consequences for society. In his 1890 monograph, he devoted an entire chapter to these matters: "The importance of climate variability for theory and practice" (see Section 4 in this chapter). In the terminology of present-day social science, Brückner transformed his academic findings into a form of "practical knowledge" (Stehr, 1991) that was meant to enable strategic responses in the economy, in the field of transportation, health care and agriculture.

He argued that the area covered by ice fields varies, the size, the water level, the appearance or even presence of lakes and rivers, the extent of floods is sensitive to climatic variations. Such disturbances would have a major impact on shipping and commercial patterns. Changing water levels and the duration of ice covers on rivers and streams, in particular, would affect the ability to navigate these waters and therefore the ease with which goods may be moved. Another most important consequence would concern agriculture more directly (see also Brückner, 1894, 1895 and our selection 6) since climatic fluctuations would have a significant influence even if the effects depend to a considerable extent on the harvested product. Brückner concluded that more than two-thirds of

[96] One year after the publication of Brückner's main work on climate variations, in 1891, a certain Professor Umlauff published a scholarly textbook on the "Foundations of Meteorology and Climatology based on most recent research." In his introduction he claims that *"literature of different people* is *linked in a mysterious manner to the climate of their homeland"* (*"So steht selbst die Literatur eines Volkes in geheimnisvollen Zusammenhängen mit den meteorologischen Elementen des von ihm bewohnten Theils des Erdballes"*) and *"Northern Europe has attained his superior level of civilization and moral because of its rain throughout the year, whereas China's success in the past was related to its summer precipitation."* (*"Nordeuropa habe es seinem Regen zu allen Jahreszeiten zu verdanken, daß es der Sitz der höchsten Gesittung wurde, so wie China seinem Sommerregen die hohe Civilisation in früher Zeit ..."*).

above-average agricultural outputs in Europe with maritime climate coincided with the warm and dry periods and an equal proportion of poor agricultural yields with the wet and cold climatic periods. In more maritime climates enhanced summer rain would cause harvests to be reduced whereas in continental climates, such as in Central North America or Russia, the summer rain would be favorable for agriculture (Brückner, 1894:2, 1915:137–138).

Thus, the two phases of the 35-year cycle would both have a beneficial effect in certain regions while disadvantages in other regions. Brückner (1915) concluded that this specific pattern of agricultural productivity change would leave its marks on the temporal variations of emigration from Western and central Europe to the United States. When conditions were favorable in Europe, namely dry and warm, *fewer* people would emigrate to the United States where a similar dryness and warmth reduced the harvests. On the other hand, when cool and wet conditions prevailed on both sides of the Atlantic, *more* people would travel across the Atlantic, because agriculture in Europe suffered from the climate while productivity in the US was increasing.

Brückner (1890:279–282) also proposed a connection between climatic fluctuations and health. He dealt with one case in some detail, namely the relationship between the appearance of typhus and the level of the ground water, which is controlled by slowly varying precipitation amounts. Having examined records of typhus mortalities in Central Europe, Brückner attributed at least part of the observed improvement in the mortality in addition to benefits derived from improvements in the sanitation system to recovering ground water levels as the result of a shift from dryer to wetter climates.

On the basis of his 35-year "mode of natural variability" and his analysis of the climate sensitivity of civilization, Brückner (1890:279,287; 1915:132) predicted a number of impending detrimental social consequences of climatic variability, in particular serious economic crises for regions that had benefited from a favorable climate in recent decades, especially areas located within the continental climate regions, such as the United States, Russia and Australia. These regions, Brückner argued, must expect an inevitable shift to dryer weather resulting in significant crop failure.

3.4.3 *The analogy to the present state of affairs*

The discussion among scientists at the turn of the century resulted in a series of findings that present-day scientists would consider to be sound and perhaps of more recent origin:

1. Climate is not constant but varies on geological as well as historical time scales.
2. Climate variability has to be differentiated between systematic, or, in Hann's words, progressive changes and temporary variations, in Hann's words, cyclical fluctuations.
3. The progressive changes were often related to human action (mainly through land-use changes, often deforestation) while temporary fluctuations were thought to be related to natural processes such as cosmic forcings and processes in the interior of the earth (including volcanic activity).

The main difference from today's discussion exists with respect to the last point. At the tum of the century, it was acknowledged that the dynamical link between climate and extra-terrestrial variations was not firmly determined. Today, most scientists are convinced that a significant part of the climate fluctuations has its origin in internal climatic processes related to the non-linearity and stochasticity of the climate dynamics.

Scientists, then as today, were confronted with a number of scientific and ethical questions. On the technical side, the problem arose of how to discriminate between human effects and internal processes. Should scientists continue with the conventional curiosity-driven research, or were the practical implications of observed patterns so serious that a purely academic orientation should be given up in favor of a more applied research orientation? Because of the perceived importance of climate for political, economic and social institutions, scientists were, and are today again, confronted with the problem whether they should merely inform or even warn society about impending climate fluctuations and demand active intervention.

The protagonists of anthropogenic climate change, or, in modern terms, environmentally conscious scientists did have an impact on the governmental-administrative level in different societies. Their message was that modifications of the environment were an agent of climate change (Grove, 1975). For instance, the American Association for the Advancement of Sciences (AAAS) in 1878 (quoted after Brückner, 1890:15) demanded reforestation programs to avoid further desiccation that was perceived to take place in North America. Demands aimed at the abatement of anthropogenic climate change were often met favorably by governments. In the 18th and 19th centuries, governmental or parliamentary committees were instituted in some countries, for instance, in Prussia, Russia, France and Italy (cf. Brückner, 1890:14–19).

A further noteworthy fact that resonates the present predicament, is the virtual silence of the "soft sciences" in the scientific and public debate. The intellectual boundaries among scientific fields hindered the incorporation of theoretical perspectives and empirical findings about climate that had been advanced in other disciplines such as the evolving social sciences. That is, the domains of the physical and the social milieu and their strict separation had already become part of the social and intellectual structure of the scientific community.

Similarly, considerable energy was then, and is now being spent by politicians, the public and others on the issue of climate change; in each case, scientists played a major role in putting the issue of climate change onto the public agenda. And politics endorsed the issue. In the past, however, the political response was mostly regional and not, as it is now demanded, global.

Brückner belonged to a small group of environmentally conscious and socially responsible scientists. He and his colleagues felt obliged to inform the public about the implications of his research. He was convinced that climate varies for natural reasons. He considered the potential for predicting these variations as a most benevolent activity since it would allow governments and social institutions to anticipate and prepare for temporary obstacles to social, political and economic developments. The task of the scientist would be to first detect the regularities and then to formulate and convey the policy options arising from this predictive capability to governments and the public at large. Aside from informing the scientific community about his results, Brückner does appear, as far as we know, to have not addressed decisions makers directly. Instead, he relied as indicated on publishing newspaper articles and presenting public lectures. However, in spite of his formidable insight into the climatological aspects, he, like some contemporaries today, overlooked that he did not have the expertise to anticipate the societal response to pronouncements about pending adverse climatic conditions, for instance by improving hygienic standards (the typhoid forecast), by perfecting the railway system (the forecast concerning the ice on the rivers) or by allowing for artificial watering of agricultural land (the forecast concerning harvests).

Hann, on the other hand, remained an academic and restricted himself to the immediate scientific problems at hand, that is, to the process of monitoring climate and understanding meteorological processes. Why he refrained from communicating more directly with the public and representatives of different social institutions, we do not know. It could be that he did not consider the results sufficiently firmly established, or he understood, unlike Brückner, that leaving one's field of expertise creates a hazardous mixture of scientific and political discourse that in the end may not be of any immediate benefit to society.

In any event, the discussion and concern about climate change quickly faded from the agenda in science and among the public in the first decades of the 20th century. We can only speculate about the reasons. Certainly, some of the practical promises associated with the new findings were found not to be fulfilled, so that the whole story would be a case of "overselling," a process later observed in the 1960s and 1970s with "rainmaking" and "cloud seeding" (Cotton and Pielke, 1992). Also, the attention was diverted to more pressing problems such as the big wars, the deep-going social repercussions and the economic disasters. Independently of the reasons, in the end a consensus emerged among climatologists (e.g., Berg, 1914:67; Lamb, 1982) that in "historical times" the global climate has been constant; that neither a warming trend nor a trend toward less precipitation takes place. Moreover, in climate science fascination with the results of new instrumental readings in the 1920s and later shifted research attention away from the issue of climate variability.

3.5 Conclusions

Our discussion of climate variability and climate change at the end of the 19th century leads to a number of conclusions that we consider relevant on methodical, theoretical and practical grounds for present-day debates:

1. The discussion about natural climate variability and anthropogenic climate change is not new. A similar debate, almost forgotten today, was going on a century ago. The protagonists found themselves in social roles and situations similar to that of contemporary scientists.
2. The early debate on the nature and consequences of climate change among climatologists, geographers and meteorologists lacked the interaction with philosophers and scientists in the emerging social sciences who had lively and vigorously written on the impact of climatic conditions on psychological and social processes for decades and centuries.
3. The attention in the academia and in the public concerning the concept of climate change and its societal implications was of limited duration. In the end, the topic lost out in the competition with other economic, political and everyday problems, and eventually also disappeared also from the research agenda of the sciences.

The specific episode we have recounted reminds us that the burgeoning genre of popularized science that surrounds present-day discussions of climate change is by no means new. Nor is it novel to acknowledge the uncertainties that surround scientific data on climate variability. It appears that the issue of climate change lends itself well to such popularization. Perhaps it does so because the issue goes to the heart of our modern common sense understanding of the natural climate as benevolent and trustworthy (cf. Stehr, 1997).

Bibliography

Arrhenius, S. A., 1896: "On the influence of carbonic acid in the air upon the temperature of the ground." *Philosophical Magazine and Journal of Science* 41: 237–276.

Arrhenius, S. A., 1903: *Lehrbuch der kosmischen Physik.* Volume 2. Leipzig: S. Hirzel.

Berg, L., 1914: "Das Problem der Klimaveränderung in geschichtlicher Zeit." *Geographische Abhandlungen* 10(2): 1–70.

Bradley, R. S., H. F. Diaz, J. K. Eischeid, P. D. Jones, P. M. Kelly, and C. M. Goodess, 1987: "Precipitation fluctuations over Northern Hemisphere land areas since the mid-19th century." *Science* 237: 171–175.

Brückner, E., 1888: "Ändert sich unser Klima?" Vortrag, Universität Dorpat (cf. *Neuen Dörptschen Zeitung* No. 68).

Brückner, E., 1889: "In wie weit ist das heutige Klima constant?" pp. 1–13 in *Verhandlungen des VIII. Deutschen Geographentages in Berlin.* Leipzig: Teubner.

Brückner, E., 1890: *Klimaschwankungen seit 1700. Nebst Bemerkungen über die Klimaschwankungen der Diluvialzeit.* Wien and Olmutz: Hölzel.

Brückner, E., 1894: "Russlands Zukunft als Getreidelieferant," pp. 1–3 in Supplement to *Münchener Allgemeine Zeitung* (November 19, 1894).

Brückner, E., 1895: "Der Einfluß der Klimaschwankungen auf die Ernteerträge und Getreidepreise in Europa." *Geographische Zeitschrift* 1: 39–51.

Brückner, E., 1902: "Zur Frage der 35jahrigen Klimaschwankungen." *Dr. A. Petermann's Mittheilungen aus Justus Perthes' Geographischer Anstalt* 48: 173–178.

Brückner, E., 1909: "Über Klimaschwankungen." *Mitteilungen der Deutschen Landwirtschafts-Gesellschaft* 24: 556–561.

Brückner, E., 1912: *Klimaschwankungen und Völkerwanderungen.* Vortrag gehalten in der feierlichen Sitzung der kaiserlichen Akademie der Wissenschaften am 13. Mai 1912. Wien: K. K. Hof- und Staatsdruckerei.

Brückner, E., [1912] 1915: "The settlement of the United States as controlled by climate and climatic oscillations," pp. 125–139 in *Memorial Volume of the Transatlantic Excursion of 1912 of the American Geographical Society.*

Brückner, E., 1923: "Julius Hann," pp. 151–160 in *Akademie der Wissenschaften in Wien, Almanach fiir das Jahr 1922.* Wien: Holder-Pichler Tempsky.

Büdel, J., 1977: *Klima-Geomorphologie.* Berlin-Stuttgart: Gebrüder Bornträger, 304 p.

Cotton, W. R. and R. A. Pielke, 1992: *Human Impacts on Weather and Climate*. Geophysical Science Series Volume 2. Ft. Collins, Colorado: ASTER Press.

DeCourny Ward, R., [1908] 1918: *Climate Considered Especially in Relation to Man*. New York and London: G.P. Putnam's Sons.

Grosjean, G., 1991: 100 Jahre Geographisches Institut der Universität Bern, *1886–1986. Jahrbuch der geographischen Gesellschaft von Bern*, 56, 1986–90, 175 p.

Grove, R. H., 1975: *Green Imperialism. Expansion, Tropical Islands Edens and the Origins of Environmentalism 1600–1860*. Cambridge University Press.

Hann, J., 1883: *Handbuch der Klimatologie*. Volume 1: Allgemeine Klimatologie. Stuttgart: J. Engelhorn.

Hann, J., 1903: *Handbook of Climatology*. Part I: General Climatology. New York: Macmillan.

Hann, J. v. and K. Knoch, [1883] 1932: *Handbuch der Klimatologie*. Fourth Edition. Volume 1: Allgemeine Klimalehre. Stuttgart: J. Engelhorn.

Hellpach, W. H., 1938: "Kultur und Klima," pp. 417–438 in Heinz Wolterek (ed.), *Klima-Welter-Mensch*. Leipzig: Quelle & Meyer.

Huntington, E., 1915: "A neglected factor in race development." *The Journal of Race Development* 6: 167–184.

Huntington, E., [1915] 1924: *Civilization and Climate*. Third Edition, Revised and Rewritten with Many New Chapters. New Haven: Yale University Press.

Huntington, E., 1916: "Climatic variations and economic cycles." *The Geographical Review* 1: 192–202.

Huntington, E., 1945: *Mainsprings of Civilization*. New York: John Wiley and Sons.

Jones, P. D., 1995: "The Instrumental Data Record: Its accuracy and use in attempts to identify the CO_2 signal," pp. 53–76 in H. von Storch and A. Navarra (eds.), *Analysis of Climate Variability: Applications of Statistical Techniques*. Berlin: Springer Verlag.

Kahlig, P., 1993: "Some aspects of Julius von Hann's contribution to modern climatology." *Interactions between Global Climate Subsystems*. The Legacy of Julius Hann. Geophysical Monograph 75.

Karl, T. R., R. G. Quayle, and P. Y. Groisman, 1993: "Detecting climate variations and change: New challenges for observing and data management systems." *Journal of Climate* 6: 1481–1494.

Knoch, K., 1932: "Vorwort zur vierten Auflage," pp. VIII–X in Julius von Hann and Karl Knoch, *Handbuch der Klimatologie*. Fourth Edition. Volume 1: Allgemeine Klimalehre. Stuttgart: J. Engelhorn.

Lacey, C. and D. Longmann, 1993: "The press and public access to the environment and development debate." *The Sociological Review* 41: 207–243.

Lamb, H. H., 1959: "Our changing climate, past and present." *Weather* 14: 299–318.

Lamb, H. H., 1982: *Climate, History and the Modern World*. London: Methuen & Co.

Le Roy Ladurie, E., [1971] 1988: *Times of Feast, Times of Famine. A History of Climate Since the Year 1000*. New York: Farrar, Straus and Giroux.

Oberhummer, E., 1927: "Eduard Brückner," pp. 195–199 in Akademie der Wissenschaften in Wien, *Almanach für das Jahr 1927*. Wien: Holder-Pichler Tempsky.

Ploetz, Al., 1911: "Die Begriffe Rasse und Gesellschaft und einige damit zusammenhangende Probleme," pp. 111–136 in *Verhandlungen des Ersten Deutschen Soziologentages vom* 13.-22. *Oktober 1910 in Frankfurt am Main*. Tübingen: J. c. B. Mohr (Paul Siebeck).

Sombart, W., [1911] 1951: *The Jews and Modern Capitalism*. Translated by M. Epstein. Glencoe, Ill.: Free Press.

Sombart, W., 1938: *Vom Menschen*. Versuch einer geisteswissenschaftlichen Anthropologie. Berlin: Buchholz & Weisswange.

Sorokin, P., 1928: *Contemporary Sociological Theories*. New York: Harper & Brothers.

Stehr, N., 1996: "The ubiquity of nature: Climate and culture." *Journal for the History of the Behavioral Sciences* 32: 151–159. Stehr, N., 1991: *Practical Knowledge*. London: Sage.

Stehr, N., 1997: "Trust and climate." *Climate Research* 8: 163–169.

Stehr, N. and H. von Storch, 1995: "The social construct of climate and climate change." *Climate Research* 5: 99–105.

Stehr, N. and H. von Storch, 1999: "An anatomy of climate determinism," in H. Kaupen-Haas (ed.), *Wissenschaftlicher Rassismus - Analysen einer Kontinuität in den Human- und Naturwissenschaften*. Frankfurt a.M., New York: Campus Verlag, 451 p.

Steinhauser, F., 1951: "Julius Hann," in Österreichische Akademie der Wissenschaften (Hrsg.), *Österreichische Naturforscher und Techniker*. Wien: Gesellschaft für Natur und Technik.

von Storch, H., and K. Hasselmann, 1996: "Climate variability and change," pp. 33–58 in G. Hempel (ed.), *The Ocean and the Poles. Grand Challenges for European Cooperation*. Jena, Stuttgart, New York: Gustav Fischer Verlag.

Whitney, J. D., 1894: "Brief discussion of the question whether changes of climate can be brought about by the agency of man, etc." *United States* Supplement I, Boston. Appendix B: 290–317.

4. Anthropogenic Climate Change: A Reason for Concern Since the 18th Century[§]

Abstract

During the last 20 years, the concept of anthropogenic climate change has left academic circles and become a major public concern. Some people consider 'global warming' as the major environmental threat to the planet. Even though mostly considered a novel threat, a look into history tells us that claims of humans deliberately or unintentionally changing climate is a frequent phenomenon in Western culture. Climate change, due to natural and anthropogenic causes, has often been dis-cussed since classical times. Environmental change including climate change was seen by some as a biblical mandate, to 'complete the Creation'. In line with this view, the prospect of climate change was considered as a promising challenge in more modern times. Only since the middle of the 20th century, has anthropogenic climate change become a menacing prospect. The concept of anthropogenic climate change seems to be deeply embedded in popular thinking, at least in Europe, which resurfaces every now and then after scientific discoveries. Also, extreme weather phenomena have in the past often been explained by adverse human interference.

A list of claims of anthropogenic climate modifications is presented; the remarkable similarity of the anthropogenic climate change debate in the second half of the 19th century is compared to the present situation. Of course, the present threat seems much more real than any of the historical predecessors, which turned out to be overestimated.

4.1 Climate change and the history of ideas

In 1890, the geographer Eduard Brückner (1890; Stehr and von Storch, 2000) wrote in his dissertation about anthropogenic climate change and natural climate variability, about winner and looser states, and about parliamentarian committees dealing with the implications of climate change:

"Very old and wide-spread is the opinion that forests have an important impact on rainfall. ... If forests enhance the amount and frequency of precipitation simply by being there, deforestation as part of agricultural expansion everywhere, must necessarily result in less rainfall and more frequent droughts. ... It is not

[§]This section originally appeared in von Storch, H. and N. Stehr: "Anthropogenic climate change: A reason for concern since the 18th century and earlier." *Geogr. Ann.*, 88A(2): 1–8, 2006.

surprising that under such circumstances the issue of a link between forests and climate has ... been addressed by governments. Lately, the Italian government has been paying special attention to reforestation in Italy and its expected improvement of the climate. ... It must be prevented that periods of heavy rainfall alternate with droughts. ... In the Unites States deforestation plays an important role as well and is seen as the cause for a reduction in rainfall. ... American Association for Advancement of Science demands decisive steps to extend woodland in order to counteract the increasing drought. ... some serious concerns. The congress for agriculture and forestry discussed the problem in detail; when the Prussian house of representatives ordered a special commission to examine a proposed law pertaining to the preservation and implementation of forests for safeguarding, it pointed out that the steady decrease in the water levels of Prussian rivers was one of the most serious consequences of deforestation only to be rectified by reforestation programs. It is worth mentioning that ... the same concerns were raised in Russia as well and governmental circles reconsidered the issue of deforestation."

Most contemporary climatologists take it for granted that the concept of anthropogenic climate change is of relatively recent origin. (By 'climate change' we do not mean changes of the local climate by the expansion of cities, clearing of single forests and other local modifications of land use. Instead, we are referring to anthropogenic changes of regional or global scale.) It is surprising for them, and most of the public, that anthropogenic climate change is by no means novel. As we can see from Brückner's discussion, concerns over extensive transformations of the earth's climate have been expressed in Europe since the 18th century Enlightenment and earlier.

Glacken (1967) offers a comprehensive analysis of Western thought about nature and culture from classical times to the end of the 18th century. The concept of human agency in changing not only climate but the environment as a whole has prevailed in Europe since classical times. Theophratus, from the 4th century BC, may be seen as a pioneer who stood at the beginning of a long history of speculation concerning climatic change and its impact on humans — climatic determinism (Stehr and von Storch, 1999). In the 18th century, the Scottish philosopher and historian David Hume (1711–1776) speculated that the recent warming could be caused by human deforestation, which would allow the rays of the sun to reach the surface of the earth. A contemporary, H. Williamson, published evidence for his view that the northern colonies of America had become more temperate in the aftermath of colonization (Williamson, 1771). Others claimed that human action would render climate more irregular and less predictable (Glacken, 1967).

People were aware of climate variations, which were evident in, for instance, the freezing of rivers, the success of harvests and the damage done by storms to dikes (e.g., Lamb, 1982; de Kraker, 1999). They speculated about the reasons for these changes, which had a significant impact on daily life. An obvious explanation was heavenly intervention, with God steering climate partly as a response to people's behavior. In Medieval times, for instance, it was proposed that climatic anomalies, or extreme events, were a punishment for parishes that were too tolerant of witches (Behringer, 1988). On the other hand, man was considered to have been placed on Earth to complete the Creation (Glacken, 1967). Environmental change was a task given to mankind by God himself. An alternative view was that Earth is organic, and would therefore age with time: "... array of occurrences seriously regarded as evidence of decay; almost any natural phenomenon was suitable: Air pollution, storminess, weather changes, earthquakes, volcanoes, and so forth" (Glacken, 1967).

4.2 History of anthropogenic climate change

In the following paragraphs, we present elements of a 'history of anthropogenic climate changes'. Most of the cases were not real; as a matter of fact, none of them proved to be associated with significant impacts related to the suggested dynamical link. But all cases were associated with the perception of significant discontinuities; in most instances, the apprehended change was seen as a threat; only rarely were they welcomed as an improvement.

1. Religious interpretations of climate anomalies, such as the prolonged wet period in England in the early 14th century, explained the adverse climatic conditions as the divine response to people's lifestyle (Stehr and von Storch, 1995). In Medieval times, for instance, it was proposed that climatic anomalies, or extreme events, were a punishment for parishes that were too tolerant of witches. Witches were believed to be able to directly cause adverse weather (Behringer, 1998). There was a sophisticated system of rogation in response to droughts in Spain (Barriendos-Vallvé and MartínVide, 1998).
2. Our oldest case documented by contemporary scientific writing refers to the climate of the North American colonies (Williamson, 1771). The physician Williamson analyzed the changes of climate, and related them to clearing of the landscape by settlers. This is a case in which human action was perceived as having a beneficial impact on climate. More cases during

Medieval times, related to colonization by monks, are described by Glacken (1967).

3. In many parts of Europe, the summer of 1816 was unusually wet, presumably because of the eruption of the volcano Tambora. However, people ascribed the adverse conditions to the new practice of using lightning conductors. The case is documented in two articles published in the newspaper *Neue Zürcher Zeitung* (21 June and 9 July 1816). The authorities called the concerns unsubstantiated and issued grave warnings concerning violent and illegal acts against the conductors. Interestingly, it is mentioned that some years earlier in Germany, people blamed the conductors for being responsible for a drought.
4. In the 19th-century scientists in Europe and in North America were confronted with the concept that the climate would be constant on historical time scales; however, scientists found significant differences between mean precipitation and temperature when averaged over different multi-year periods (e.g., Brückner, 1890). Also, scientists claimed that the water levels of rivers would fall continuously. This led to questioning of the assumption of constant climatic conditions — in modern terms, interdecadal natural variability — and, alternatively, to the hypothesis that the observed changes are caused by human activities, mainly deforestation or reforestation. It seems that the majority adopted the concept of man-made causes over the natural variability hypothesis (Brückner, 1890; Stehr *et al.*, 1996).
5. The developing science of forestry (e.g., Grove, 1975) informed people that the severe floods in Switzerland in the middle of the 19th century would be related to logging of forests in the high mountains (Pfister and Brändli, 1999). The flooding was falsely perceived as novel so that a novel cause was sought; as a consequence, the Swiss Forest Law was instituted, an early progressive environmental law, which was most useful to limit an unsustainable practice, but was based on false scientific arguments.
6. There are reports that both the extensive gun-fire during the First World War and the initiation of short-wave trans-Atlantic radio communication were blamed for wet summers in the 1910s and 1920s (Kempton *et al.*, 1995; Hinzpeter, pers. comm.).
7. In the first part of the 20th century, a remarkable warming took place in many parts of the world. In 1933, this warming was documented, and the uneasy question 'Is the climate changing?' was put forward in *Monthly Weather Review* (Kincer, 1933). Some years later, Callendar (1938) related the warming to human emissions of carbon dioxide into the atmosphere, a mechanism described some 40 years earlier by Arrhenius (1898). Interestingly, Arrhenius (1903) himself stated that anthropogenic emissions of CO_2 would cause a

significant climate change only after several hundred years. Flohn (1941) also brought this line of reasoning into the scientific debate. In the 1940s, global mean temperatures began to fall, which eventually led to claims that Earth was heading towards a new Ice Age.

8. After the Second World War scientists noticed a cooling and some speculated whether this cooling was the first indication of a new Ice Age, possibly brought on by human actions, mostly emissions of dust and industrial pollution. It was speculated that human pollution would increase by a factor of as much as eight which could increase the opacity of the atmosphere within 100 years by 400%. This in turn would significantly reduce incoming sunlight causing the global mean temperature to sink by 3.5 K. Such a cooling would almost certainly be enough to force Earth into a new Ice Age (Rasool and Schneider, 1971). The prospect was illustrated with the words: 'Between 1880 and 1950 Earth's climate was the warmest it has been in five thousand years. ... It was a time of optimism. ... The optimism has shriveled in the first chill of the cooling. Since the 1940s winters have become subtly longer, rains less dependable, storms more frequent throughout the world.' (Ponte, 1974:89).
9. After the Second World War, the new practice of exploding nuclear devices in the atmosphere caused widespread concern about the climatic implications of these experiments. According to Kempton's analysis, even nowadays many lay-people are concerned about this link (Kempton *et al.*, 1995).
10. In Russia, plans for re-routing Siberian rivers southward have been discussed since the beginning of the 20th century. The plans visualize benefits in supplying semi-arid regions with water, and an improved regional climate. A by-product was thought to be an ice-free Arctic Ocean because of the reduced freshwater input from the rivers. This would shorten the winters and extend the growing season; the increase of evaporation from the open water would transform the Arctic climate into a maritime climate with moderate temperatures and busy harbors along the Soviet Union's north coast (Ponte, 1974:136). Such plans were formally adopted in 1976 at the 25th Assembly of the Soviet Communist Party. Scientists from the West as well as from the Soviet Union opposed these plans and warned that the formation of an ice-free Arctic could significantly affect the global ocean circulation and thus global climate. Eventually, the plans were abandoned although more careful analyses indicated that the probability of melting the Arctic sea ice associated with a re-routing of the rivers was overestimated (e.g., Lemke, 1987).
11. The concept of engineering or manipulating the climate system became popular in the first half of the 20th century. Re-routing Siberian rivers was one such scheme; another was put forward by the New York engineer Riker, who in

1912 suggested changing the Gulf Stream with the purpose of improving the climate not only of North America but also of the Arctic and Europe. Riker's idea was summarized by Ponte (1974:138):

> *The Gulf Stream travels up along the American coast without any problem, but when it turns east to cross the Atlantic Ocean it collides with the Icy Labrador Current coming down from the Arctic. This collision in relatively shallow water weakens the Gulf Stream ... But this would change ... if a simple jetty 200 miles long could be built from Cape Race on Newfoundland to a point just beyond the underwater Grand Banks. The jetty would keep the two currents apart ... Off the tip of Greenland ... the more powerful Gulf Stream would divide. Half would throw increased warmth against Northern Europe, and half would thrust into the Arctic... The benefits of this would be enormous. Fog would disappear, all ice in the Arctic would melt. The melting of the Arctic would improve the world climate in two ways.... Europe and North America would be freed of chilling storms and icy ocean currents... And without the North Polar ice, the surviving ice pack at the South Pole would become the heaviest part of our planet. Centrifugal force would then tip the Earth ... With the Northern hemisphere tipped more towards the sun, Europe and North America could expect warmer climate.*

It is interesting to note that Riker thought of warming as an improvement of climate. The same view is put forward by H. Lamb (1982). Also, the idea of modifying ocean currents was later pursued by scientists from the USA, USSR and other nations. In most cases, these schemes revolved around the building of a dam, which would, for instance, block the flow through the Bering Strait. Ponte (1974) offered a sketch displaying a large variety of such plans (Fig. 1).

12. Similar to the idea of climate engineering is the military use of climate modifications. The idea of changing the course of the Gulf Stream had been put forward in the 18th century by Benjamin Franklin, who envisaged a northward diversion of the Gulf Stream as a powerful weapon against the British Empire (Ponte, 1974:137). A perceived attack using climate as a weapon is a purported Soviet plan in the 1950s to build a "jetty 50 miles or more long out from near the eastern tip of Siberia. The jetty would contain several atomic-powered pumping stations that would push cold Arctic waters down through the Bering Strait. This would … inject increasing amounts of icy waters into the ocean current that flows down the west coast of Canada and the United States. The result would be colder, more stormy weather throughout North America and enormous losses to the American economy in agriculture, work days and storm damage" (Ponte, 1974:169–170). Concern about the development of

Alternative II: Change Weather and Climate

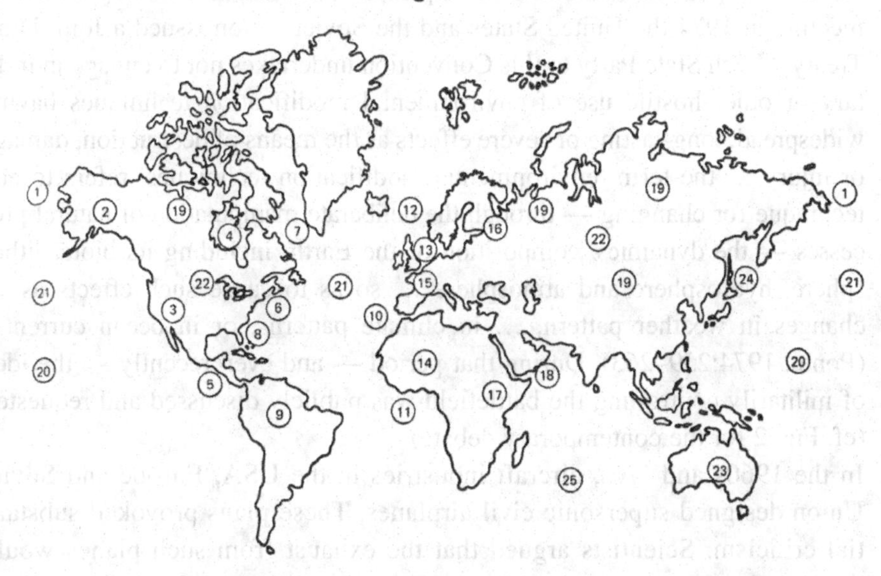

1. Dam Bering Strait
2. Dam Yukon River to create giant lake
3. Create "Lake Fallacy" in Arizona
4. Artificially heat Hudson Bay or dam its southern tip at James Bay
5. Blast sea level canal across Central America
6. Dam Long Island Sound
7. Dam Labrador Current
8. Dam Gulf Stream at Bimini Strait
9. Dam Amazon River to create inland sea
10. Dam Strait of Gibraltar
11. Use heating or other means to increase evaporation in or near Gulf of Guinea
12. Dam or blast submarine ridges near Norwegian Straits
13. Dam English Channel
14. Divert ocean or river waters to create inland sea in Sahara
15. Dig canal from North Sea to Mediterranean Sea
16. Dig canal from Arctic Ocean to Baltic Sea
17. Dig canal to eliminate swamps from upper Nile River in Sudan
18. Dam Red Sea
19. Divert Arctic-flowing rivers southward or dam them
20. Cloud-seed the trade winds
21. Heat or cool spots of ocean surfaces to alter global winds
22. Put stationary dust clouds over enemy nation, either in atmosphere or in outer space
23. Dig canal to ocean to turn Lake Eyre into inland sea
24. Dam Tatar Strait between Sakhalin Island and Siberia
25. Tow icebergs from Antarctica.

Figure 1: Locations and short description of various plans to modify environmental conditions in order to change climate (from Ponte, 1974).

climate weapons led to a series of diplomatic discussions. During a summit meeting in 1974 the United States and the Soviet Union issued a Joint Draft Treaty: "Each State Party to this Convention undertakes not to engage in military or other hostile use of environmental modification techniques having widespread, long-lasting or severe effects as the means of destruction, damage or injury ... the term 'environmental modification techniques' refers to any technique for changing — through the deliberate manipulation of natural processes — the dynamics, composition of the Earth, including its biota, lithosphere, hydrosphere and atmosphere ... so as to cause such effects as ... changes in weather pattern, ... in climate patterns, or in ocean currents" (Ponte, 1974:259–263). During that period — and even recently — the idea of militarily controlling the battlefield was publicly discussed and requested (cf. Fig. 2 for the contemporary debate).

13. In the 1960s and '70s, aircraft industries in the USA, Europe and Soviet Union designed supersonic civil airplanes. These plans provoked substantial criticism. Scientists argued that the exhaust from such planes would damage the ozone layer in the stratosphere and the climate in general. In the USA the plans were stopped, but in Europe, the Concorde was built and in the Soviet Union the TU 144. Of course, numerous military supersonic aircraft are nowadays cruising the lower stratosphere. For many years, discussion about the impact of air traffic on the climate ceased. But in the early 1990s, the topic re-entered the public debate, this time regarding high-flying conventional jetliners. The focus of concern is the effect of contrails and exhaust gases on the radiative balance of Earth. Scientists (e.g., Sausen and Schumann, 1998) regard present effects from these sources as minor compared to other effects. However, some argue that with present projections of future passenger numbers and technology the effect could be significant.
14. A popular, but for natural scientists somewhat surprising, mechanism links space traffic to a deteriorating global climate. In Kempton *et al.*'s (1995) interviews with lay-people, this mechanism is mentioned several times: 43% of the respondents in Kempton's survey considered the statement "there may be a link between the changes in the weather and all the rockets they have fired into outer space" plausible.
15. The ongoing deforestation of tropical forests is of great concern to many people, who are afraid not only of reduction in the variety of species but also of changes in global climate (Kempton *et al.*, 1995; Dunlap *et al.*, 1993). Model calculations indicate that these land-use modifications cause significant local and regional changes whereas in most model calculations global effects are marginal. Interestingly, similar results were obtained for the

Figure 2: Public perspectives of future weather modification possibilities.

climatic implications of the transformation of the North American wilderness into agricultural land (Copeland *et al.*, 1996).

16. Anthropogenic aerosols are considered powerful agents for changing the global climate. One scenario deals with the emission of aerosols mainly from burning forests and fossil fuels. A dramatic version is that of 'nuclear winter'

in which it was assumed that the explosion of a multitude of nuclear bombs in a future war would create a high-flying veil of soot particles which would effectively shut off solar radiation and cause a collapse of the biosphere (Cotton and Pielke, 1995). Support came from a number of computer simulations. The ignition of the Kuwait oil wells in the aftermath of the 1991 Gulf War led some scientists to expect a minor nuclear winter, particularly with respect to the Indian monsoon. It turned out that the effect was severe locally but insignificant on the larger scale (e.g., Cahalan, 1992).

17. A new line of concern, especially in Europe, refers to the stability of the Gulf Stream in the Atlantic Ocean. Ocean models exhibit a markedly nonlinear behavior of the Atlantic circulation with two stable states, one with an active Gulf Stream and another with a weakened northward transport moderating the European climate. Both states are stable within a certain range of conditions, but when the system is brought to the margins of these ranges, it can switch abruptly to the other state (Marotzke, 1990). Palaeoclimatic reconstructions using evidence from ice cores and other indirect sources support the existence of such stable states and frequent rapid changes from one state to another. In the global warming debate the risk of a 'collapse' of the Gulf Stream is put forward. While the globe is becoming warmer, Europe and northeast America would experience colder conditions.

4.3 Social and cultural processes

Which social and cultural processes make the concept of anthropogenic climate change not merely an episodic but an almost permanent issue that challenges scientists and alarms non-expert? Under present circumstances, such social processes include the need for scientists to frame their problems so that they fit the area of their expertise, the readiness of members of the scientific community to engage in public agenda setting, and the desire of scientists to have a presence in the media (Bray and von Storch, 1999b). The fact that concern about climate has prevailed not only in recent decades but for many centuries indicates is indicative that humans depend fundamentally on the reliability of climate, and that sometimes this reliability is perceived as being endangered. It is interesting to note that climate change mostly takes an apocalyptic form, with the appearance of extremes, more severe droughts and floods, and more violent storms (Glacken, 1967; Ponte, 1976). We suggest that anthropogenic climate change is of permanent, often dormant concern for people in the West. It can be revived at any time by weather extremes, which are, at least in modern times, not taken as rare but normal events, but as scripture on the wall spelling imminent homemade disaster. In former times, the attention ceased after a while when conditions returned to normal.

Because of the open, complex character of the climate system (Oreskes *et al.*, 1994), the long-time scales involved and homogeneity problems of the observational record, knowledge about the climate system will always suffer from significant uncertainty. On the other hand, as we have seen, climate change is an important topic, arousing public interest and concern. Thus, climate change research is bound to be post-normal science (Funtowicz and Ravetz, 1985), characterized by high uncertainty and high stakes (Bray and von Storch, 1999a), with policy and science influencing each other, and public, antagonistic debates among not only scientists but activists and other non-experts as well.

In most of the instances we have listed, the actual threat of anthropogenic climate change was either absent or an extravagant claim made by the scientific community. Of course, in the present case of 'global warming', we do not know at this time if it is a real threat or if the warnings are exaggerated as in earlier cases. The fact that the Intergovernmental Panel on Climate Change (IPCC; e.g., Houghton *et al.*, 2001) is examining the scientific evidence with great care and in 1995 made its famous statement that "the balance of evidence suggests that there is a discernible human influence on global climate" and that other official bodies such as the Enquete Commission of the Deutscher Bundestag (1988) voiced grave concerns, may be considered as support for the reality of the envisioned threat.

However, 100 years ago, parliaments and governments in Europe (e.g., Prussia, Italy and Russia) also established distinguished committees that were asked to deal with the reality of anthropogenic climate change related to deforestation (Brückner, 1890). And about 200 years ago the British Parliament was discussing the climatic implications of human modifications in British tropical colonies (Grove, 1975).

The notion that we are really facing large-scale man-made climate at this time is supported by a large body of scientific analysis, of which possibly the so-called 'detection and attribution' -studies are the most important (IDAG, 2005). 'Detection' means that the record of observational evidence is examined if the most recent changes are beyond the range of natural variability (Hasselmann, 1993). For temperature, at least, this condition is fulfilled, so that it is concluded that non-natural factors are at work. In the next 'attribution' step (Hasselmann, 1998), climate model simulations are screened for that mix of responses to various anthropogenic factors which best describes the most recent trend: This is a mix of elevated atmospheric greenhouse gas and aerosol concentrations. Thus, there is good empirical evidence that anthropogenic climate change is taking place now (see also Houghton *et al.*, 2001).

4.4 Conclusions and outlook

So far, climate research has almost exclusively been concerned with the natural-science dimension, that is, with the dynamics of the climate system, its sensitivity to external disturbances and perspectives for the foreseeable future. This section points out that there is also a cultural dimension of 'climate'. Only a few researchers are actively engaged in studying the social and cultural processes of speaking about climate, of the formation and usage of lay knowledge, of the formation and social functioning of mental images, icons and popular explanations of climate and its interaction with people. We are in need of social and cultural sciences to map, understand and, as far as possible, predict the social and cultural construction of climate. These issues are not only of academic interest but have a significant bearing on the ongoing public debate about anthropogenic climate change (von Storch and Stehr, 2000). This type of knowledge is urgently needed to guide policy makers and the public in developing and adopting rational policies for dealing with the very real prospect of significant future climate change (Sarewitz and Pielke, 2000).

References

Arrhenius, S.A., 1896: On the influence of carbonic acid in the air upon the temperature of the ground. *Philosophical Magazine and Journal of Science*, 41: 237–276.

Arrhenius, S.A., 1903: *Lehrbuch der kosmischen Physik*. Volume Two. S. Hirzel, Leipzig, 1026 p.

Barriendos-Vallvé, M. and Martín-Vide, J., 1998: Secular climatic oscillations as indicated by catastrophic floods in the Spanish Mediterranean coast area (14th–19th centuries). *Climatic Change*, 38: 473–491.

Behringer, W., 1988: *Hexenverfolgungen in Bayern*. R. Oldenbourg Verlag, München, 546 p.

Bray, D. and von Storch, H., 1999a: Climate Science. An empirical example of postnormal science. *Bulletin of the American Meteorological Society*, 80: 439–456.

Bray, D. and von Storch, H., 1999b: Climate Science and the transfer of knowledge to public and political realms. In: von Storch, H. and Flöser, G. (eds.), *Anthropogenic Climate Change*. Springer Verlag, Berlin, pp. 287–328.

Brückner, E., 1890: *Klimaschwankungen seit 1700 nebst Bemerkungen über die Klimaschwankungen der Diluvialzeit. Geographische Abhandlungen herausgegeben von Prof. Dr. Albrecht Penck in Wien*. E.D. Hölzel, Wien and Olmütz, 325 p.

Cahalan, R., 1992: Kuwait oil fires as seen by Landsat. *Journal of Geophysical Research*, 97: 14565–14571.

Callendar, G.S., 1938: The artificial production of carbon dioxide and its influence on temperature. *Quarterly Journal of the Royal Meteorological Society*, 64: 223–239.

Copeland, J.H., Pielke, R.A. and Kittel, T.G.F., 1996: Potential climatic impacts of vegetation change: A regional modeling study. *Journal of Geophysical Research*, 101(D3): 7409–7418.

Cotton, W.R. and Pielke, R.A., 1992: *Human Impacts on Weather and Climate*. ASTeR Press, Ft. Collins, 288 p.

De Kraker, A.M.J., 1999: A method to assess the impact of high tides, storms and storm surges as vital elements in climate history. The case of stormy weather and dikes in the Northern part of Flanders, 1488–1609. *Climatic Change*, 43: 287–302.

Deutscher Bundestag, 1988: *Schutz der Erdatmosphäre: Eine internationale Herausforderung*. Deutscher Bundestag, Referat Öffentlichkeitsarbeit, Bonn, 582 p.

Dunlap, R.E., Gallup, G.H. Jr., and Gallup, A.M., 1993: Health of the Planet: A George H. Gallup Memorial Survey. Gallup International Institute, Princeton, New Jersey, USA.

Fleming, J.R., 1998: *Historical Perspectives on Climate Change*. Oxford University Press, Oxford, 194 p.

Flohn, H., 1941: Die Tätigkeit des Menschen als Klimafaktor. *Zeitschrift für Erdkunde*, 9: 13–22.

Funtowicz, S.O. and Ravetz, J.R., 1985: Three types of risk assessment: A methodological analysis. In: Whipple, C. and Covello, V.T. (eds.), *Risk Analysis in the Private Sector*. Plenum, New York, pp. 217–231.

Glacken, C.J., 1967: *Traces on the Rhodian Shore*. University of California Press, 763 p.

Grove, R.H., 1975: *Green Imperialism. Expansion, Tropical Islands Edens and the Origins of Environmentalism 1600–1860*. Cambridge University Press, 540 p.

Hasselmann, K., 1993: Optimal fingerprints for the detection of time dependent climate change. *Journal of Climate*, 6: 1957–1971.

Hasselmann, K., 1998: Conventional and Bayesian approach to climate change detection and attribution. *Quarterly Journal of the Royal Meteorological Society*, 124: 2541–2565.

Houghton, J.T., Ding, Y., Griggs, D.J., Noguer, M., van der Linden, P.J., Dai, X., Maskell, K., and Johnson, C.A., 2001: *Climate Change 2001: The Scientific Basis*. Cambridge University Press, 881 p.

IDAG, 2005: Detecting and attributing external influences on the climate system. A review of recent advances. *Journal of Climate*, 18: 1291–1314.

Kempton, W., Boster, S.J., and Hartley, J.A., 1995: *Environmental values in American Culture*. MIT Press, Cambridge, MA and London, 320 p.

Kincer, J.B., 1933: Is our Climate Changing? A Study of long-term temperature trends. *Monthly Weather Review*, 61: 251–259.

Lamb, H.H., 1982: *Climate, History and the Modern World*. Methuen, London, 387 p.

Lemke, P., 1987: A coupled one-dimensional sea ice-ocean model. *Journal of Geophysical Research*, 92(C12): 13164–13172.

Marotzke, J., 1990: Instabilities and multiple equilibria of the thermohaline circulation. PhD thesis, Universität Kiel, 194 p.

Oreskes, N., Shrader-Frechette, K., and Beltz, K., 1994: Verification, validation, and confirmation of numerical models in earth sciences. *Science*, 263: 641–646.

Pfister, C. and Brändli, D., 1999: Rodungen im Gebirge — Überschwemmungen im Vorland: Ein Deutungsmuster macht Karriere. In: Sieferle, R.P. and Greunigener, H. (eds.), *NaturBilder*. Wahrnehmungen von Natur und Umwelt in der Geschichte. Campus Verlag, Frankfurt/New York, pp. 9–18.

Ponte, L., 1976: *The Cooling*. Prentice-Hall, Englewood Cliffs, NY, 306 p.

Rasool, S.I. and Schneider, S.H., 1971: Atmospheric carbon dioxide and aerosols: Effects of large increases on global climate. *Science*, 173: 138–141.

Sarewitz, D. and Pielke, R., Jr., 2000: Breaking the global-warming gridlock. *The Atlantic Monthly*, July 200: 55–64.

Sausen, R. and Schumann, U., 1998: Estimates of the climate response to aircraft emission scenarios. Institut für Physik der Atmosphäre 95, DLR, 26 p.

Stehr, N. and von Storch, H., 1995: The social construct of climate and climate change. *Climate Research*, 5: 99–105.

Stehr, N. and von Storch, H., 1999: An anatomy of climate determinism. In: Kaupen-Haas, H. (ed.), *Wissenschaftlicher Rassismus — Analysen einer Kontinuität in den Human- und Naturwissenschaften*. Campus Verlag, Frankfurt and New York, 451 p.

Stehr, N. and von Storch, H. (eds.), 2000: *Eduard Brückner — The Sources and Consequences of Climate Change and Climate Variability in Historical Times*. Kluwer, Dordrecht, 338 p.

Stehr, N., von Storch, H., and Flügel, M., 1996: The 19th century discussion of climate variability and climate change: Analogies for present day debate? *World Research Review*, 7: 589–604

von Storch, H. and Stehr, N., 2000: Climate change in perspective. Our concerns about global warming have an age-old resonance. *Nature*, 405: 615.

Williamson, H., 1770: An attempt to account for the change of climate, which has been observed in the Middle Colonies in North America. *Transactions of the American Philosophical Society*, 1: 272.

Chapter 4

Cultures of Science

The two sections in this chapter are attempting to diagnose the state and complementary roles of natural and social sciences in climate research. In the last decades, the natural sciences, and specifically physics, dominate in informing the public discourse and the policymaking, while social sciences only play a marginal role (Section 1). A comparison of the different cultures of physical and social scientific thinking in Section 2 reveals significant differences in their virtues and potentials. In Section 3, we examine, together with theoretical physicist Armin Bunde, the limitations of physical sciences in advising policymaking.

1. Climate Protection*

It is not space but the structuring that comes from the soul that has social significance.

Georg Simmel ([1908] 1992)

Soil and climate together determine the natural fertility of a country and of its people who are led either to indolence or to activity.

Werner Sombart (1938)

Abstract

The voice of the social sciences in climate research and climate policy discussions, except for interventions from economists mainly about the costs associated with policy options driven by climate science research, has been muted if not altogether absent. The absence of the social sciences from climate research and policy not surprisingly has colored climate discourse in peculiar ways. We are making the case for a greater involvement and importance of the social sciences in interdisciplinary climate research.

Keywords: Social science, Climate change, Adaptation, Mitigation, Climate protection

1.1 Introduction

Throughout much of their history, the social sciences have been torn; as the quotes from Georg Simmel and Werner Sombart demonstrate between those who advocate either incorporating "nature" into social science discourse or dislodging any reference whatsoever to natural forces from social science. It is evident that contemporary social science discourse has generally ruled out environmental or physical (as well as biological) factors as directly relevant to sociological, economic, historical or anthropological "explanations." There are good reasons that account for the differentiation of cognitive agendas in science, chief among them are the following:

*This section originally appeared in Stehr, N, and H. von Storch: "Climate protection," *Journal of Consumer Protection and Food Safety* 4: 56–60, 2009. DOI: 10.1007/s00003-008-0392-y.

— biological and cultural evolution are not identical,
— the natural environment of society is for the most part independent of human action,
— societies have succeeded in emancipating themselves from many environmental constraints.

Nonetheless, the ecosystem, refashioned to a lesser or greater extent by social action by way of appropriating its resources, remains a major material source and constraint for human conduct. Social scientists today have, for the most part, accepted the firm dichotomy of nature and society. The social sciences have their own distinct domain of inquiry, their own methods and theories: A world of objects and subjects that constitutes therefore a reality sui generis.

The upshot of these intellectual developments in social science has been that the voice of the social sciences in climate research and in climate policy discussions has been muted if not altogether absent, except for interventions from economists, mainly about the costs associated with policy options driven by climate science research.

In the following brief remarks about "climate protection," we would like to show how the absence of the imagination of the social sciences from climate research and policy discussion sustains in scientific and political discussion about global climate change, a singular focus on mitigation efforts in response to the threat of global warming. We begin with the case of tropical diseases that are widely anticipated to move poleward and that are seen to constitute one of the major health risks associated with climate change. The threat of tropical diseases moving into regions of the world now mostly unaffected by such health hazards is often used to make the case the reduction of emissions, i.e., dealing with the cause of anthropogenic climate change would be the only meaningful approach. This argument neglects the fact that better adaptive measures have made many areas free of such diseases, to begin with (Reiter, 2001). From there, we move directly to the case for a stronger emphasis both in research and in policy devoted to measures designed to protect societies from a changing climate. This discussion to date has also been singularly dominated by approaches favored by the natural sciences.

1.2 Tropical diseases and social conduct

As is well known, there are many warnings that certain diseases currently concentrated in the world's tropical regions, such as malaria or dengue fever, will wander poleward and threaten to become a widespread danger to the health of the population in the temperate world. However, as many have now noted, this is an unlikely scenario. Socioeconomic factors are far more significant than climate in the determination of disease prevalence. For example, there are one thousand times more cases of dengue in the Northern regions of Mexico than in Southern Texas (Gubler *et al.*, 2001). The climate across this 100-km band is the same, even the vector habitats are similar in many instances, but the pattern of social interactions and access to public health are vastly different. Socializing outside at dusk, when the mosquitoes quest for food, is prevalent in Mexico. North of the border, the people are indoor in air-conditioned rooms, socializing around a television set. Air conditioning is an adaptive measure, just like sitting outside in the cool evening air. One exposes the public to a health hazard, the other can protect them from vector-borne diseases as well as heat waves. Adaptation is how humans have come to occupy so many different climate conditions around the world. Proactive adaptation is our best chance to keep people from harm. Disease and disability tend to be obscured by a climatic focus, so too are social influences and circumstances. Yet, these are often the most powerful determinants of health. Inequality and poverty kill (Kawachi *et al.*, 1997; Marchand *et al.*, 1998).

What this case in our view demonstrates is that the absence of social sciences discourse from climate research leads to extremely simplified and allegedly causal linkages between changes in the environment. The same conclusion applies as far as health risks that are seen to be associated with climate according to a study carried out by Brikowski *et al.* (2008). The authors claim to have discovered a direct linkage between a significant increase in the incidence of nephrolithiasis (a kidney stone disease) and warming in the United States. What is completely missing from their inquiry as well as consideration is how such changes in their impact on society are mediated by social conduct. If the latter would be taken into account, the analysis of health risks and possible climate policies would have looked very different.

1.3 Climate protection

We would like to introduce our observations on climate change and climate policies in the form of mitigation and adaptation strategies by explicating a peculiar compound noun from contemporary German political discourse on climate, namely the term *Klimaschutz* (in this instance: measures to protect the climate or climate protection). The concept *Klimaschutz* exemplifies the peculiar dilemma we would like to explore next in this paper. If we are not mistaken, in its current virtually taken-for-granted meaning, *Klimaschutz* means protecting climate (and the environment) from society. But it could just as well mean to insulate society from climate change (and associated environmental change), especially if the concept is supposed to refer to "global warming management."

The policy of climate protection, with the support of influential circles within climate research, is predominantly one-sided. It is not the appropriate way to deal with the problem. Up to now, climate protection policies rely almost exclusively on measures that have to do with energy, transport, industry and housing infrastructure: such as measures to save energy and to increase efficiency, and the corresponding legislative frameworks. Often conferences about regional climate issues are in reality conferences about local exercises to save energy, without any reference to local aspects of ongoing and possible local future climate change. Such a policy is hardly concerned with future sewage and rainwater management, or with implementing measures to counteract previous urban warming as a means to mitigate the local manifestation of global warming (Gill *et al.*, 2007), or with the dangers of storm surges related to tropical cyclones such as Nargis earlier this year in Myanmar.

The threat posed to the basic living conditions of society by climatic changes cannot be combated, as it has been up to now, only by protecting the climate from society, particularly given that many of these measures are of a symbolic nature. Additional effective efforts are required on the part of researchers, politicians and economic leaders in order to come to terms with the climatic dangers that already exist today, some of which will intensify in the future, even in the face of a successful climate protection policy. This protection cannot wait to be put in place only after we have lived through catastrophes in the wake of weather extremes such as 2008's devastating storm surge in Myanmar caused by the tropical storm Nargis; rather, it must be realized in the form of precautionary measures. And these are in short supply here and now!

Sometimes such a proposal is countered with the declaration that extending the existing climate protection policy by means of an active precautionary climate policy is essentially identical with admitting that the existing policies have miscarried. This argument is obviously short-sighted and unfounded.

1.4 A reduction of CO_2 emissions is insufficient

Concentrating climate policy on the reduction of greenhouse gases serves no purpose if it leads at the same time to preventing taking precautions in dealing with present dangers and their possible future amplifications. Such a one-sided research perspective and climate protection policy will neither protect the climate from society in the coming decades nor society from the climate.

In contrast, our conception faces up to reality and its demands: Climatic warming is not a fleeting, temporary or short-lived phenomenon. It is important to state this outright, because the impression is often given, intentionally or otherwise, that the climate can be changed in one direction or the other in a short span of time.

Lowering emissions means, in the first place, only reducing the increase in their concentration. And, in fact, it would already be a triumph if we were presently able to reduce the increase of these emissions. The long-term prevention of global warming, however, requires a quite extensive reduction of greenhouse gas emissions, i.e., lowering human emissions to almost zero. The length of time necessary for our elevated concentration of CO_2 to return even approximately to its original — here, preindustrial — equilibrium amounts to somewhere between several decades and a few centuries.

Why are these time spans relevant? On the one hand, they point to the prodigious efforts that are necessary worldwide in order to effectively halt climatic warming; on the other, these numbers are the point of departure for our further theses regarding how society will have to deal with the consequences of climatic warming.

1.5 Reducing energy consumption does not reduce risks

Adaptation and prevention, i.e., reduction of emissions, are reasonable options that must be pursued in concert. As a rule, however, they are different options. Adaptation to the dangers posed by the climate will only incidentally reduce emissions; likewise, energy-saving and other reductive measures will only seldom be able to reduce the vulnerability of our basic living conditions in face of the dangers posed by the climate. What both options have in common, however, is that they are promoted by means of technological innovations, but most particularly by means of social changes. A realistic assessment and a public discussion of the dangers of climate and climate change are the first prerequisites for understanding the nature and the extent of the social changes required. A positive atmosphere, in which innovations are actively promoted and publicly acknowledged, is useful not only in the context of an active climate policy.

1.6 Climate policies should follow a dual strategy

Reductive measures limit the extent of anthropogenic climate change and they push for the efficient usage of limited resources, even only part of them. Furthermore, they represent an engine for general technological progress. Thus, they are in any case reasonable and necessary. Adaptive measures, on the other hand, are also in any case reasonable and necessary, which allow to deal with the unavoidable vagaries of present climate and possible future climates. They also propel technological progress.

The two sorts of human response to the perspective of anthropogenic climate change are not contradictions, but an inseparable pair of twins. Reduction of emissions is a long-term project, while measures lead to visible impacts much faster, but these effects will be beneficial also when the reductive measures begin to work at a later point in time. The more effective the reduction, the more efficacious the adaptive measures — in the long term!

Often it is said that pursuing adaptation would lead to the illusion that the necessary level of reduction of emissions would not really be needed; that considering adaptation would imply belittling of the dangers of climate change — that adaptation and mitigation would indeed be exclusive alternatives — either adaptation or mitigation. Interestingly, this argument is not brought forward by those who demand thinking and planning about adaptation now, but by those who insist that the only real option is energy saving, and everything else would be a mere distraction. When listening to advocates of this sort, sometimes one gets the impression that they are not really interested in dealing with the dangers of climate but that they just use the specter of climate change as a leverage to impose a rebuilding of economy and society to their liking.

We concentrate in this paper on the issue of adaptation, not because it would be more important than efforts to reduce emissions but because it is not receiving the attention it needs.

1.7 The dangers of a one-sided approach to climate change

Let us proceed in a thought experiment from the premise that human beings on this planet could manage to meet the goal of reducing emissions by 80% in the space of one year. When, under these conditions, would the climate machine achieve a new "equilibrium"? The answer is: Not for decades. In other words, the climatic change that is already underway cannot be prevented overnight, even by the greatest imaginable efforts in the realm of mitigation policy.

A climate policy that commits itself to the problem of mitigation while neglecting the urgent need for adaptation is an irresponsible climate policy, because it denies society's inevitably higher degree of vulnerability in the coming decades. The goal of such a policy — to protect the climate from society and thereby to protect society from itself — will bear fruit only in the distant future.

A representative example of the prevailing one-sidedness of the discussion of climate protection and efforts in this area is the often dispassionately employed term "heat deaths." As if people were almost inevitably and defencelessly victims of nature, and not victims of specific social circumstances; and indeed of social circumstances that irresponsibly put people at the mercy of extreme heat and its consequences, and do not preventively shield the segments of the population that are most severely affected (Klinenberg, 2002). To speak of "heat deaths," as was done in the case of the hot summer of 2003, protects only the municipalities, regions or countries that failed in their duty to take precautions. The very use of this term guarantees, so to speak, that the trends that are the actual cause of this phenomenon will be thoughtlessly repeated.

1.8 A carbon-free world arrives too late

There are at least three important reasons why politicians, society and scientists must urgently think in terms not only of mitigation but also of precautionary measures, as a reaction to the consequences of climate change.

The time scales of the long-term results of lowering emissions and of climate change do not correspond to each other. Any successes in terms of reducing the emission of greenhouse gases will take effect, as we have said, only in the far future. A world in which only small amounts of CO_2 are still being emitted will come too late to limit climate change in the next decades. The practically unlimited emissions of the past and up to now guarantee that climate change will change our future living conditions. The dilemma lies in the fact that the time scales of nature are not congruent with those of political decision-making cycles in democratic societies, which proceed in terms of election periods and cycles of attention, and which are reflected in the limited horizons of human action.

The threat posed by extreme climatic events, such as torrential rains, floods and heat waves, is already considerable today, and always has been in many regions of the world. One need only to recall New Orleans in 2005; the storm surge of 1872 on the German Baltic coast or that of 1953 in Holland; or even Hurricane Mitch, which was turned to good use in the course of the 1992 negotiations in Rio de Janeiro. The vulnerability of our basic living conditions increases parallel to the growth of the global population in endangered regions, where growing segments of the population are marginalized without protection and, not least for reasons of political economy, become victims of extreme weather events.

The regions of the world whose basic living conditions will be particularly hard hit by the consequences of worldwide climatic changes are already demanding today, rightfully and increasingly vehemently, that the world must see to their protection, and not only to the protection of the climate.

1.9 The Kyoto approach has failed

Worldwide climate policy, like that of Germany as well, is particularly clearly represented by the Kyoto Protocol. The Kyoto Process concerns itself almost exclusively with questions of reduction. The reduction targets of the Kyoto Protocol, which expires in 2012, will hardly be achieved. The successful execution of the Kyoto Protocol's so-called "Clean Development Mechanism" (CDM), in terms of the worldwide emission of CO_2, would by 2012 reduce the volume of worldwide cumulative emissions only little compared to the same development without Kyoto reductions (cf. Wigley, 1998).

For developing and emerging countries, particularly China and India, there is currently no obligation to reduce greenhouse gas emissions. Their share of the global balance of greenhouse gases is continually increasing. At least in the near future, the developed societies will also emit more climate-damaging greenhouse gases. The total emission of CO_2 above all, despite all efforts at reduction, will probably increase further in industrialized countries between now and 2012.

The Kyoto approach, as a form of socially restrictive, large-scale global planning, has failed (Prins and Rayner, 2007). Any subsequent process based on this hegemonic planning mentality will serve no purpose. As a result, climate change of human origin is steadily advancing and will step up in the future. A reversal of this alteration to our global climate will be possible only over the span of decades, if not centuries.

1.10 Prevention has a higher legitimacy

Despite the contrary opinions of all political parties up to now and their reluctance to speak publicly about precautionary climate programs, adaptation as a precautionary measure is relatively easy to implement and to legitimize in political terms. Moreover, it has the enormous advantage that its success will be evident in the foreseeable future. When it comes to finding solutions to a problem by means of innovations in science and technology, it is easier to present these in the form of adaptive measures.

1.11 Adaptation is regional

The consequences of warming vary significantly according to region and climatic zone. Research into precautionary measures thus means expanding our knowledge about regional changes. To what, exactly, are we going to have to adapt? With the aid of adaptive strategies, several goals at once can be achieved, because they are primarily locally or regionally oriented, and therefore can be flexibly configured: Improving quality of life, decreasing social inequity and increasing political participation are not mutually exclusive.

The dual challenge of adaptation and prevention also leads to a reasonable division of labor. The German federal and European responsibility falls at the level of the frameworks for managing emissions, while for those in charge of sub-national regions and municipalities, the question of reducing their vulnerability should have priority. In fact, institutions and persons charged with specific responsibilities — for coastal protection or for the Hamburg harbor, for instance — demonstrate a concrete commitment to solving problems of adaptation.

1.12 The failures of virtuous conduct

In the public discussion, down to the present day, only reduction activities have been portrayed as a virtuous form of behavior, even when it merely takes the form of purely symbolic and largely ineffective actions, such as Sundays without driving, doing without long trips, or staging public events. This perception is not unproblematic, to the extent that it gives actors the impression that sufficient steps are being taken to protect the climate and society.

A revision or extension of this perception to include a proactive attitude toward precautions and toward necessary social changes, however, as is essential to protect society from the changing climate and thus to reduce the vulnerability of the very basis of our existence, is still lacking. An effective defense of this basis demands precautionary measures in the coming years and decades. This must now be our priority.

The prospects that such a research program will be seen to be important and ultimately will be carried even in countries where the resistance in policy and research toward adaptation strategies is strong, such as in Germany, should be enhanced considerably if social scientists reconsider the firmly established boundaries between the domains of inquiry of the natural and the social sciences. Even more generally, both in the case of adaptation research/policy advice and mitigation research/policies, the territory should not be left solely to the natural sciences.

References

Brikowski, T. H., Lotan, Y., and Pearle, M. S. (2008) Climate-related increase in the prevalence of urolithiasis in the United States. *Proc Natl Acad Sci* 105:9841–9846.

Gill, S. E., Handley, J. F., Ennos, A. R., and Paulett, S. (2007) Adapting cities for climate change: The role of the green infrastructure. *Built Environ* 33:115–133.

Gubler, D. J., Reiter, P., Ebi, K. L., Yap, W., Nasci, R., and Platz, J. A. (2001) Climate variability and change in the United States: Potential impacts on vector- and rodent-borne diseases. *Environ Health Perspect* 109:223–233.

Kawachi, I., Kennedy, B. P., Lochner, K., and Prothrow-Smith, D. (1997) Social capital, income inequality, and mortality. *Am J Publ Health* 87:1491–1498.

Klinenberg, E. (2002) *Heat Wave. A Social Autopsy of Disaster in Chicago*. Chicago: The University of Chicago Press.

Marchand, S., Wikler, D., and Landesman, B. (1998) Class, health, and justice. *The Milbank Quarterly* 76:449–467.

Prins, G. and Rayner, S. (2007) The wrong trousers. Radically rethinking climate policy. James Martin Institute for Science and Civilisation, Oxford, 37 p.

Reiter, P. (2001) Climate change and mosquito-borne disease. *Environ Health Perspect* 109(Suppl. 1):141–146.

Simmel, G. ([1908] 1992) *Soziologie*. Volume 11 of the Gesamtausgabe. Suhrkamp: Frankfurt am Main.

Sombart, W. (1938) *Vom Menschen*. Versuch einer geisteswissenschaftlichen Anthropologie. Berlin: Buchholz & Weisswange.

Wigley, T. M. L. (1998) The Kyoto Protocol: CO_2, CH_4 and climate implications. *Geophys Res Lett* 25:2285–2288.

2. Micro/Macro and Soft/Hard: Diverging and Converging Issues in the Physical and Social Sciences†

Abstract

The concept of scales is widely used in social, ecological and physical sciences, and is embedded in various ongoing philosophical debates about the nature of nature and the nature of society. The question is whether the difference between scales makes a difference and if so what difference. Multilevel approaches compete with reductionist approaches. We are tracing the highlights of the disputes as well as some of the resolutions that have been offered. Most importantly, debates about differences in scale are enmeshed in what should be distinguished, namely analytical knowledge-guiding interests and those that might be called practical knowledge-guiding interests. It is unlikely that purely analytical debates can be resolved. However, progress about the impact and relevance of scale can be achieved with respect to the practical-political discursive level of knowledge claims. More specifically, scales are a crucial concept in determining the capacity for action from knowledge about the dynamics and structures of processes. For instance, in the context of climate change, knowledge claims about global and continental processes are relevant for the international political process aimed at abatement measures, whereas knowledge about regional and local effects controls decisions concerning adaptation measures.

2.1 Introduction and overview

Climate scientists share a greater common understanding of the scientific usefulness of scales than do social scientists. This greater agreement among climate scientists does not necessarily enhance the practicality of the knowledge claims about the dynamics of the climate system. Social scientists have debated the relevance of different scales for a long time, and though the arguments have been rehashed and repeated many times, they have rarely led to new insights. Conflicts gave way to a search for linkages between micro and macro levels of analysis and the failure to agree on linkages-reanimated conflicts (cf. Alexander and Giesen, 1987). The disputes remain unresolved. We will try to reframe the issue rather than repeating claims that are invariably contested.

For the purpose of further reflection, the main points we want to develop in the process of reframing the debate on scaling is that scales — or the difference between micro and macro, as many social scientists would say — are relevant not

---

†This section originally appeared in Stehr, N. and H. von Storch: "Micro/macro and soft/hard: Diverging and converging issues in the physical and social sciences," *Integrated Assessment* 3: 115–121, 2002.

just as an analytical problem (i.e., as a problem of scientific description or explanation) but as a practical problem.

The disputes about scale have rarely been treated as a topic that ought to distinguish between knowledge-guiding interests that are concerned, on the one hand, with the *practicality of the knowledge* generated by science and, on the other hand, with optimizing certain *theoretical and methodological conceptions* in the process of generating knowledge claims (see Gibson *et al.*, 1998:14).

The practicality of knowledge generated by science refers to the usefulness knowledge may have as a "capacity for action" in practical circumstances and for particular actors. Analytical attributes of knowledge refer to methodological and theoretical attributes of knowledge claims, e.g., the extent to which propositions developed for one level can be generalized to another level or the extent to which they can be formalized. The practicality of knowledge claims, in contrast, aims to assist actors, confronted with specific conditions of action, to set something into motion and do so, of course, with the aid of knowledge.

We maintain that there is not a linear relation or obvious congruence between enhancing the analytical and practical capacity of knowledge. Two examples may illustrate the point.

1. The determination that the "growing division of labor in society explains the rising divorce rates in advanced society" constitutes a prominent and eminent social science explanation. However, a nation, a region, a city, a village or a neighborhood will hardly be able to "manipulate" the division of labor and therefore "arrest" (in the sense of effect) divorce rates within its boundaries.
2. The insight that the equilibrium global temperature of Earth would rise by, say, 2°C if CO_2 concentrations in the atmosphere double does not provide people at the regional and local level with the capacity to react skillfully, as this insight on the global scale provides no assessment for ongoing environmental change on a regional or local scale within the foreseeable future.

Knowledge-guiding interests that aim to enhance the practicality of knowledge claims and knowledge claims that live up to specific analytical attributes (such as logic, truthfulness, reality-congruence, etc.) are not mutually exclusive; however, they do not necessarily lead to identical knowledge claims.

The distinction between analytical and practical is particularly relevant to actors who have to deal with and convert scientific knowledge claims into practical action. Thus, choices of scale not only affect what can or will be analyzed but also what can or will be done.

But first, we need to restate and summarize the social and the physical science debate about the role of scales in the analysis and the differences that are claimed on behalf of a differentiation with the help of scales. In the case of physical science, our description will focus on climate science.

2.2 Scales in the social sciences: Mixing levels or what is the difference?

In every living thing what we call the parts is so inseparable from the whole that the parts can only be understood in the whole, and we can neither make the parts the measure of the whole nor the whole the measure of the parts; and this is why living creatures, even the most restricted, have something about them that we cannot quite grasp and have to describe as infinite or partaking of infinity.

Johann Wolfgang von Goethe (1785)

Goethe maintains that the understanding of parts or wholes requires the elimination of their difference. It appears that the social sciences have generally followed his advice since a liberal mixing of levels or multilevel analysis is common in social science accounts. Even in approaches that are self-consciously micro or macro, linkages between levels are evident. If this is the case, then the difference between levels is unnecessary.

The assertion whether a differentiation is helpful or not is based on a certain comprehension of the constitution of examined processes and therefore to specific knowledge-guiding interests internal to the scientific community. For example, the common theoretical link that sociologists obtain between the conduct of individual actors (micro level), situational factors or the social structure typically is a particular social psychological theory (macro level). When Robert K. Merton (1938) explains deviant behavior, he does so not as the outcome of individual differences but as the consequence of the situation within which the actor is located. Merton argues that unattainable goals produce deviant behavior. Whether the actor in fact faces unattainable goals is determined by the situation or social structure. Situations vary, but the social psychology that links actor and situation (namely, trying to pursue legitimate goals) is the same for each individual. Hence the differences in location explain deviance. Without the social psychological premises, the account would be incomplete (Zelditch, 1991:102–103). Put another way, the problem is that neither solitary perspective "pays adequate attention to the *constructed* nature of both individuals and groups" (Calhoun, 1991:59). Part and system form a whole. The mixture of different scales is argued to be constitutive for social phenomena. Paraphrasing Wittgenstein ([1953] 1967:20,20e), understanding parts of an ordinary language game requires the comprehension of a form of life or a cultural system.

As the label already indicates, the institutionalist perspective assigns explanatory priority to the macro scale: "Social processes and social change ... result at least in part, from the actions and interactions among large-scale actors ... Welfare systems, job markets, and cultural structures become products of organizations or sets of organizations" (Meyer, 1987:17). Network analysis, rational choice theory, interaction ritual chain analysis (Collins, 1981) or Homans' (1961) behaviorism typically favor the micro scale. These strategies simply maintain and are linked to the theoretical premise that the realities of social structure reveal patterns of "repetitive micro-interaction" (Collins, 1981:985).

What is relevant and constitutes the immediate environment for the analysis depends on prioritizing scales. Macro models — where their own internal divisions of levels are problematic — prefer resource- or ecological-dependency perspectives, while micro models that acknowledge the presence of levels emphasize cultural practices and conceptions as their most relevant environment.

Approaches that readily acknowledge and freely mix different scales in their analysis place different emphasis on relevant scales, on how one progresses down or up the conceptual scale (aggregation, cumulation and interaction), and on how robust or recalcitrant different units of analysis happen to be.

The strict limitation to certain scales, i.e., the conviction that levels cannot be mixed, is based on considerations of methods or access to levels. As Scheff (1990:27–28) states in an exemplary fashion: The macroworld, "so vast and so slow moving, requires special techniques to make its regularities visible — the statistics and mathematical models now taken for granted. The study of the microworld also requires special techniques, but for the opposite reason: the movements are too small and quick to be readily observable to the unaided eye." Our interpretation of the elevation of one level is one necessitated by perspective: The perspective of the observer as compared with the level of the observer.

The debate about levels of analysis in the social sciences is not constrained or disciplined by commonly accepted definitions of the boundaries of disciplines and subdisciplines. However, the choice to work within the accepted confines of subatomic physics or cellular biology a priori limits the resolution of patterns that can legitimately be studied. Social scientists have not reconstructed the world of social phenomena in the same hierarchical fashion that is generally taken for granted in the physical sciences.

2.3 Scales in the physical sciences: The climate system

A characteristic of the physical climate system is the presence of processes on all spatial scales. The "scale" of a process is the extension of an area where the direct impact of the process is felt. Thus, the spatial scale of the tropical trade wind

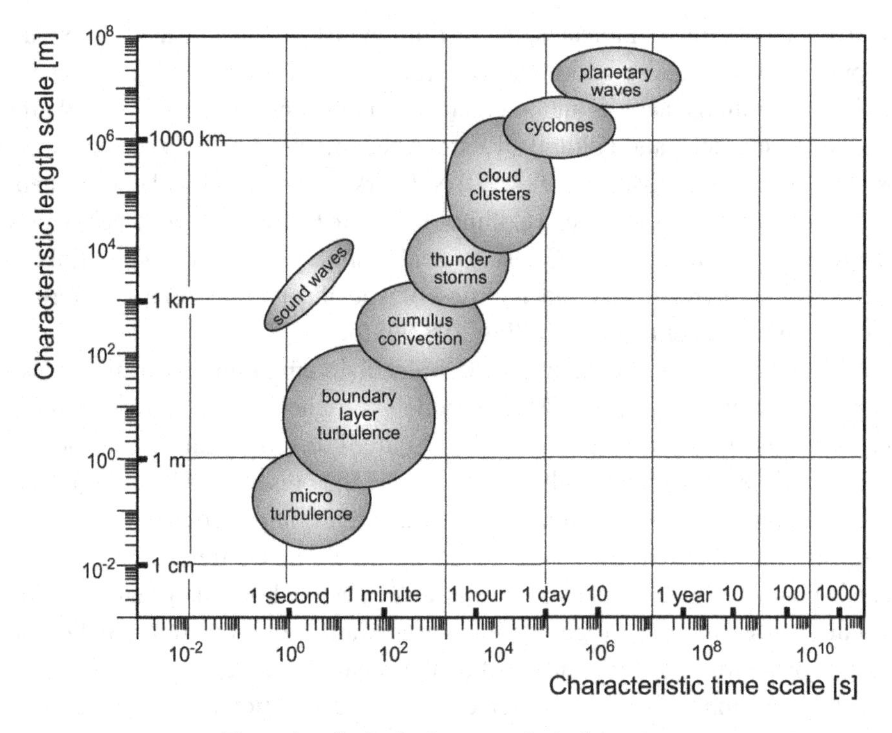

Figure 1: Scales in the atmospheric dynamics.

system is several thousand kilometers; that of a cyclone at mid-latitudes is about one thousand kilometers; a front, a few hundred kilometers; a thunderstorm, a few kilometers; and individual turbulent eddies in the atmospheric boundary layer exert an influence on scales of several meters and less (Fig. 1). A typical feature of this cascade of spatial scales is that it is associated with a similar cascade in temporal scales. Smaller scales exhibit shorter-term variations, whereas larger scales vary on longer time scales. For instance, a cyclone with a diameter of a thousand kilometers exists for several days, whereas a thunderstorm of several kilometers' diameter is dissipated after a few hours (Fig. 1). A similar analysis can be made for oceanic processes.

All of these processes interact. The trade wind system, as part of the Hadley Cell, helps to maintain a meridional temperature gradient at mid-latitudes so that the air flow becomes unstable and eddies form (namely, extratropical cyclones); these storms form fronts, and the strong winds blowing above the Earth surface create a turbulent boundary layer of several hundred meters height. In this argument, large-scale features create environmental conditions so that smaller-scale features emerge. This view is supported by an experiment with a complex climate model simulating atmospheric motion on an "aqua planet," i.e., a globe without topography (Fischer *et al.*, 1991). Initiated with a motionless state, driven by equator-to-pole gradients in the global ocean's surface temperature and by solar radiation, the general circulation of the atmosphere just described emerges within a few weeks, with trade winds, extratropical storms and turbulent boundary layers. Climate at a smaller scale appears as conditioned by the state at a larger scale (von Storch, 1999).

However, the smaller scale is not determined by the larger scale, as demonstrated by the weather details, which may differ greatly in two very similar synoptic situations (Starr, 1942; Roebber and Bosart, 1998). But information about the conditioning large-scale state is incorporated in the statistics of small-scale features. This fact is used in paleoclimatic reconstructions (Appenzeller *et al.*, 1998; Mann *et al.*, 1998), which are based entirely on "upscaling" of local information like tree ring widths or densities.

Do the smaller scales affect the larger scales? They do: Without the small-scale eddies in the turbulent boundary layer, a cyclone would not lose its kinetic energy; without the extratropical storms, a much stronger equator-to-pole temperature gradient would appear and the Hadley Cell, with its trade wind system, would possibly extend to the polar regions. While the large scales condition the smaller scales, the smaller scales make the large scales more fuzzy. There is a simple intuitive argument for this asymmetry: There are many realizations of the smaller-scale process, encompassed in the area of influence of one larger-scale process. The smaller-scale processes represent a random sample of possible realizations, and their feedback on the large-scale process depends on the statistics of the smaller-scale processes. The details of a single storm are not relevant, but the preferred area of formation, the track of the storms and the mean intensity do influence the formation of the general atmospheric circulation.

Aside from making the large scales more fuzzy, smaller-scale short-term variations also cause the large-scale components to exhibit slow variations. This phenomenon, comparable with Brownian motion of macroscopic particles under the bombardment of infinitely many microscopic molecules, is demonstrated in the "stochastic climate model" of Hasselmann (1976). The short-term variations are considered random, and the large-scale components integrate this random behavior. Whether the many small-scale features are really varying randomly is irrelevant; as long as these processes are strongly nonlinear, often a valid assumption, their joint effect cannot be distinguished from randomly generated numbers.

This effect is illustrated in Fig. 2, showing the time evolution of a one-dimensional world characterized by a large-scale (global) temperature: Solar (short-wave) radiation is intercepted by this world; part of this radiation is

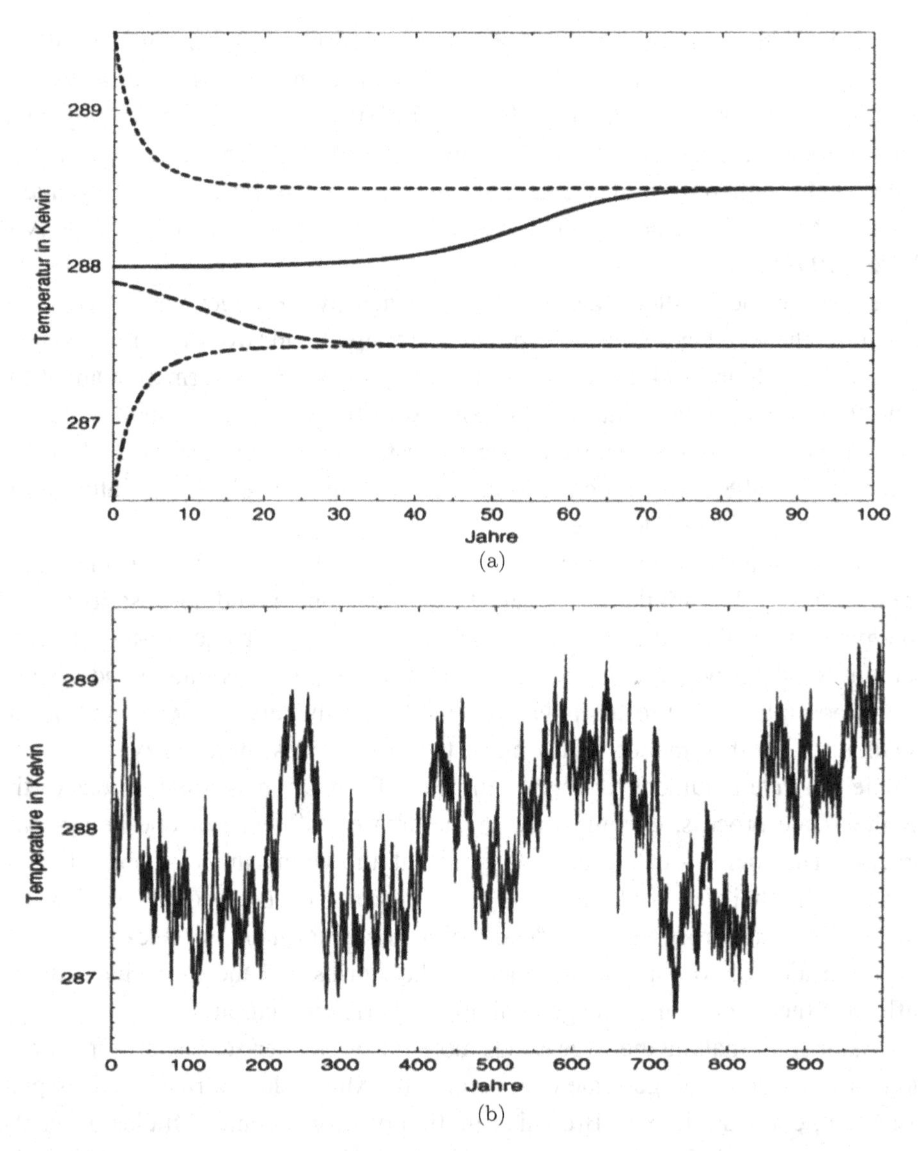

Figure 2: Performance of energy-balanced model: (a) when "no noise" is present and (b) if it is activated.

reflected back to space; the intercepted radiation is re-emitted as thermal (long-wave) radiation proportional to the fourth power of temperature. When the proportion of reflected solar radiation ("albedo") is such that a higher temperature is connected with lower reflectivity (less snow and ice) and lower temperature with higher reflectivity (more snow and ice), then Earth can have two different temperatures. Which of these temperatures is attained depends on where one starts (Fig. 2(a)). However, a different behavior emerges when the reflectivity exhibits additional random variations, representing the variable small-scale cloud cover of Earth (Fig. 2(b)). The systems exhibit slow variations and intermittent jumps between the two preferred regimes of the system. Obviously, in this thought experiment, the small-scale, short-term variations ("noise") are a constitutive element, causing the emergence of slow variations of large-scale temperature (von Storch *et al.*, 2001). Time series of observed large-scale quantities, like the global mean near-surface temperature, show similar frequency behavior, even if the interesting regime shifts in Fig. 2(b) are not obvious (Hansen and Sutera, 1986; Nitsche *et al.*, 1994).

2.4 There is nothing as practical as a good theory

Our discussion of the macro/micro controversy in the social sciences and the accomplishments of scaling in climate science have shown that, despite their divergence, the focus in both cultures is on the *analytical* accomplishments. That is, scaling issues tend to be deliberated and judged in the sciences based on the internal knowledge-guiding interests.

But this also implies that the scaling problem is discussed in a one-sided manner. Improvements in the analytical capacities of knowledge (or the scientificity of knowledge claims) do not always improve upon the practical efficacy of knowledge. The thesis that analytical improvements enhance the usefulness of knowledge is best captured in the maxim "*there is nothing as practical as a good theory*." The emphasis clearly is on *good* theory, and what constitutes good theory is disputed more in the social than the physical sciences. An improvement of theory surely constitutes intellectual progress within science. But good theory does not invariably point to "elements" in a concrete situation that can be acted upon in order to accomplish a certain purpose, e.g., in the sense of affecting development of a specific process — even though that process is better understood because of the good theory (and the scaling choices made in order to generate good theory).

That good theory — and whatever good theory may mean in concrete terms — does not automatically mean practical knowledge can best be shown by defining *knowledge as a capacity to act or as a model for reality* (see Stehr, 2000).

Our choice of terms is inspired by Francis Bacon's famous observation "scientia est potentia," or, as it has often been somewhat misleadingly translated: "knowledge is power." Bacon suggests that knowledge derives its utility from its capacity to set something in motion. The term "potentia," or capacity, describes the power of knowing. Human knowledge represents the capacity to act, to set a process in motion or to produce something. The success of human action can be gauged from changes that have taken place in reality or are perceived by society.

The notion of knowledge as a capacity for social action has the advantage that it enables one to stress not just one dimension, but the rich, multifaceted consequences of knowledge for action. The realization of knowledge in political, everyday, economic or business contexts is embedded in a web of social, legal, economic and political circumstances. That is, the definition of knowledge as a capacity for action strongly indicates that the realization of knowledge is dependent on specific social and intellectual contexts. Knowledge use and its practical efficacy are a function of "local" conditions and contexts.

Scaling decisions can therefore be affected with respect to actionable circumstances and not merely attributes that suggest themselves because they happen to be desirable from an analytical perspective.

2.5 The differences that make a difference: Scales in climate change and climate impact research

The scale problem outlined above relates to both a success and a major limitation of modern climate research in constructing plausible climate change scenarios. The computing technology available now and in the foreseeable future does not allow resolution of small-scale features in climate models. Instead, the small-scale features are not described in any detail but are parameterized, i.e., their effect on the resolved scales is described as a function of the resolved scales. In this way, the equations are closed, and the large-scale features are described realistically. The overall general circulation of the atmosphere is simulated as in the real world, extratropical storms are formed with the right life cycles and locations. Obviously, this success is not perfect, and the next years will see significant improvements. Independently of the degree of success on scales of, say, 2,000 km and more, today's global climate models fail to provide skillful assessments on scales of say 100 and less kilometers.

Therefore, the contemporary discussion concentrates only on anthropogenic climate change detectable now on the global scale, and not on the regional and local scale. For political purposes, namely for emphasizing the need for abatement action of the world's governments, these results valid for large scales are sufficient, as the details of expected change are less important than the perception of global risk.

When we consider the alternative, though not contradictory political strategy to abatement measures, namely adaptation, we need regional and local assessment of anthropogenic climate change since climate impacts people mainly on the regional scales. Regional scales as social constructs are highly variable. Storm surges happen regionally; the storm track may be shifted by a few hundred kilometers; when rain replaces snowfall or snow melts early, a catchment is affected, and so on. Such information may be derived by postprocessing the output of global climate models, by exploiting the above-sketched links between the scales. For this purpose, climate scientists have designed dynamically or empirically constructed models describing the possible regional states consistent with large-scale states generated in global models. This approach is named "downscaling," as information from larger scales is transferred to smaller scales. "Dynamical downscaling" uses models based on detailed dynamical models or regional climate models; "empirical downscaling" operates with statistical models fitted to the observational evidence available from the recent history.

While a large variety of "downscaling" techniques has been developed in the past decade, they have not yet provided climate impact research with the required robust estimates of plausible regional and local climate change scenarios, mainly because global climate models have not yet provided sufficiently converged consistent large-scale information to be processed through "downscaling" (Giorgi *et al.*, 2001). However, one might expect that this gap could be filled in within a few years so that detailed regional and local impact studies may provide robust scenarios of changes in climatic variables like temperature, storminess and sea level.

This information also has to be postprocessed further with dynamical and empirical models of climate-sensitive systems, like the water balance in a catchment, the ecology of a forest, the statistics of waves on marginal seas or the economy of agriculture. Of course, in many cases, this postprocessing is futile if other factors are considered in parallel to changing climatic conditions, such as changing social preferences, technological progress and the like.

These models again suffer from scale problems. Almost all environmental modeling efforts assume that the system may be separated into two subsystems, one that is explicitly described and another that is considered noise, which influences the explicitly described part statistically. The explicitly described "dynamical" part is considered to carry the essential dynamics. In climate and other physical systems, the dynamical subsystem comprises all large-scale processes while the noise subsystem comprises the small-scale processes. Thus, the former contains relatively few processes and the latter, infinitely many. This convenient separation according to scales can no longer be adopted in other systems, such as ecosystems or economies.

2.6 Conclusions

In the physical sciences, discussions of scale revolve around time and place. In the social sciences, discussions of micro/macro tend to concentrate on functional relationships. The concepts of macro vs. micro and of scales in the social and in the physical science are widely used, but not without problems (see Connolly, 1983:10–44). The question is whether the difference between scales makes a difference, and if the scales matter, what difference they make. Not surprisingly, the intensity of the dispute varies by discursive field. In the physical sciences, in this case, climate science, the debate is less intense and manifests itself in more definitive knowledge claims about the impact of differences in scale.

Well-intentioned scientists focus on the analytical qualities of the knowledge claims they generate, largely because they see it as the solution to the question of "what is to be done," without looking at how effective and practical these accounts are going to be. This can be judged to be a form of escape from scientific labor. Effectiveness and practicality are governed by prevailing social conditions. The ability to transform prevailing contexts requires, first, an examination and identification of those contextual elements that can be altered. The mutable conditions then drive decisions about scaling.

References

Alexander, J. C. and B. Giesen, 1987. "From reduction to linkage: The long view of the micro-macro debate." In J. C. Alexander *et al.* (eds), *The Micro-Macro Link*. Berkeley: University of California Press, pp. 1–42.

Appenzeller, C., T. F. Stocker, and M. Anklin, 1998. "North Atlantic Oscillation dynamics recorded in Greenland ice cores." *Science* 282:446–449.

Bauman, Z., 1990. *Thinking Sociologically*. Oxford: Blackwell.

Calhoun, C., 1991. "The problem of identity in collective actions." In J. Huber (ed.), *Macro-Micro Linkages in Sociology*. Newbury Park, California: Sage, pp. 51–75.

Collins, R., 1981. "The microfoundations of macrosociology." *American Journal of Sociology* 86:984–1014.

Connolly, W. E., 1983. *The Terms of Political Discourse*. Princeton, New Jersey: Princeton University Press.

Fischer, G., E. Kirk, and R. Podzun, 1991. "Physikalische Diagnose eines numerischen Experiments zur Entwicklung der grossräumigen atmosphärischen Zirkulation auf einem Aquaplaneten." *Meteorogische Rundschau* 43:33–42.

Gibson, C., E. Ostrom, and T.-K. Ahn, 1998. *Scaling Issues in the Social Sciences*. A Report for the International Human Dimensions Programme on Global Environmental Change. IHDP Working Paper 1. Bonn: IHDP.

Giddens, A., 1990. "R.K. Merton on structural analysis." In J. Clark, C. Modgil and S. Modgil (eds.), *Robert K. Merton: Consensus and Controversy*. London: Falmer Press, pp. 97–110.

Giorgi, F., B. Hewitson, J. Christensen, M. Hulme, H. von Storch, P. Whetton, R. Jones, L. Mearns, and C. Fu, 2001. "Regional climate information — evaluation and projections." In J. T. Houghton *et al.* (eds.), *Climate Change 2001. The Scientific Basis*. Cambridge University Press, pp. 583–638.

Hansen, A. R. and A. Sutera, 1986. "On the probability density function of planetary scale atmospheric wave amplitude." *Journal of the Atmospheric Sciences* 43: 3250–3265.

Hasselmann, K., 1976. "Stochastic climate models. Part I. Theory." *Tellus* 28:473–485.

Homans, G. C., 1961. *Social Behavior: Its Elementary Forms*. New York: Harcourt, Brace, and World.

Huber, J. (ed.), 1991. *Macro-Micro Linkages in Sociology*. Newbury Park, California: Sage.

Knorr-Cetina, K. and A. V. Cicourel (eds.), 1982. *Advances in Social Theory and Methodology: Toward and Integration of Micro and Macro-Sociologies*. London: Routledge and Kegan Paul.

Mann, M., R. S. Bradley and M. K. Hughes, 1998. "Global-scale temperature patterns and climate forcing over the past centuries." *Nature* 392:779–789.

Merton, R. K., 1975. "Social knowledge and public policy. Sociological perspectives on four presidential commissions." In M. Komarovsky (ed.), *Sociology and Public Policy. The Case of Presidential Commissions*. New York: Elsevier, pp. 153–177.

Meyer, J., F. O. Ramirez, and J. Boli, 1987. "Ontology and rationalization in the Western cultural account." In G. M. Thomas *et al.* (eds), *Institutional Structure. Constituting State, Society and the Individual*. Newbury Park, California: Sage, pp. 12–37.

Münch, R. and N. J. Smelser, 1987. "A theory of social movements, social classes, and castes." In J. C. Alexander *et al.* (eds), *The Micro-Macro Link*. Berkeley: University of California Press, pp. 371–386, 403 and 404.

Nitsche, G., J. M. Wallace, and C. Kooperberg, 1994. "Is there evidence of multiple equilibria in the planetary-wave amplitude?" *Journal of the Atmospheric Sciences* 51:314–322.

Parsons, T., 1954. *Essays in Sociological Theory*. New York: Free Press.

Pedlosky, J., 1987. *Geophysical Fluid Dynamics*. New York: Springer.

Roebber, P. J. and L. F. Bosart, 1998. "The sensitivity of precipitation to circulation details. Part I: An analysis of regional analogs." *Monthly Weather Review* 126:437–455.

Scheff, T. J., 1990. *Microsociology. Discourse, Emotion and Social Structure*. Chicago: University of Chicago Press.

Starr, V. P., 1942. *Basic Principles of Weather Forecasting*. New York: Harper.

Stehr, N., 1992. *Practical Knowledge*. London: Sage.

Stehr, N., 2000. *Knowledge and Economic Conduct: The Social Foundations of the Modern Economy*. Toronto: University of Toronto Press.

von Storch, H., 1999. "The global and regional climate system." In H. von Storch and G. Flöser (eds.), *Anthropogenic Climate Change*. New York: Springer Verlag.

von Storch, H., S. Güss and M. Heimann, 1999. *Das Klimasystem und seine Modellierung. Eine Einführung.* New York: Springer.

von Storch, H., J.-S. von Storch, and P. Müller, 2001. "Noise in the climate system — Ubiquitous, constitutive and concealing." In B. Engquist and W. Schmid (eds.), *Mathematics Unlimited — 2001. and Beyond.* New York: Springer, pp. 1179–1194.

Wittgenstein, L., [1953] 1967. *Philosophical Investigations*. New York: Macmillan.

Zelditch, M. Jr., 1991. "Levels in the logic of macro-historical explanations." In J. Huber (ed.), *Macro-Micro Linkages in Sociology*. Newbury Park, California: Sage, pp. 101–106.

3. The Physical Sciences and Climate Politics[‡]

3.1 Orientation

In the following sections, the physic-ness of climate science is discussed. One of the motivations for doing so is related to the observation that some scientists, trained as physicists, often play a very influential role as political actors, when interpreting and explaining the significance of scientific insights for policy implications using the "linear" model, according to which knowledge leads directly to first political consensus and then decision making, while social and cultural processes related to preferences and values represent mostly invalid disturbances (e.g., Beck, 2010; Curry and Webster, 2011). Of course, the linear model means that those in control of the knowledge ought to be in control of the outcome of the political decision process.

We therefore thought it meaningful to examine to what extent this claim of political competence is warranted. We find that it is not. In order to become societally relevant, climate science has to become trans-disciplinary, by incorporating the social-cultural dimension.

We could also have done an analysis with fields related to ecology or economics. A similar phenomenon is also observed among some high-profile members of these groups; those scientists find it difficult to balance the authority of scientific competence, limitations and integrity with the need to engage one's own values. We limit ourselves here to physics, first because this field is likely the most important.

Our chapter features three main sections.

In Section 3.2, we discuss the historical development of the concept of climate leading us from an anthropocentric view to a strictly physical world view, and one that is now moving once again towards a more anthropocentric view — this time concerning not only the impacts but also the drivers. This is not meant as a general review of the history of climate sciences, which is done competently by Weart (2011). Instead, we want to emphasize the circularity in the development, from an anthropocentric view, over an impassionate, distanced truly physical view, back to an anthropocentric view.

In Section 3.3, a series of physical issues, from modeling, over parameterizations, the impossibility of experimentation, and data problems is discussed.

‡This section originally appeared in von Storch, H., A. Bunde, and N. Stehr: "The physical sciences and climate politics," in J.S. Dyzek, D. Schlosberg, and R.B. Norgaard (eds.), *The Oxford Handbook of Climate Change and Society*. Oxford University Press, Oxford, UK, 2011, pp. 113–128.

In Section 3.4, the concept of "post-normal" science is introduced, which is related to high uncertainties in the field of climate research, and the high stakes on the societal side. Here, at the boundary between science and policy, new dynamics emerge, which have little to do with physics; dynamics which depend on culture and history, on conflicting interests and world views.

A brief concluding section argues for the need for a trans-disciplinary approach to climate in order to assist in developing policies consistent with physical insights and cultural and social constraints.

3.2 History of climate science

Historically, "climate" was considered part of the human environment. Alexander von Humboldt ([1845] 1864:323–324) in 1845 in his book *Cosmos: A Sketch of a Physical Description of the Universe* defined climate as the sum of physical influences, brought upon humans through the atmosphere:

> *The term climate, taken in its most general sense, indicates all the changes in the atmosphere, which sensibly affect our organs, as temperature, humidity, variations in the barometrical pressure, the calm state of the air or the action of opposite winds, the amount of electric tension, the purity of the atmosphere or its admixture with more or less noxious gaseous exhalations, and, finally, the degree of ordinary transparency and clearness of the sky, which is not only important with respect to the increased radiation from the earth, the organic development of plants, and the ripening of fruits, but also with reference to its influence on the feelings and mental condition of men.*

Thus, like astronomy, climate in much of 19th-century discourse was subject to an anthropocentric view. The global climate was little more than the sum of regional climates (cf. Hann, 1903), and the challenge was to faithfully describe regional climates by measuring and mapping the statistics of their weather. Not surprisingly, a large body of information was generated, dealing with the impacts of climate on people and societies. It was the time of the prominence of the perspective of climatic determinism (Fleming, 1998; Stehr and von Storch, 1999, 2010). At the turn of the 19th and 20th centuries, questions were formulated more in terms of climate as a physical system (e.g., Friedmann, 1989; see also the systematic approach presented by Arrhenius, 1908), and meteorology and oceanography became "physics of the atmosphere" and "physics of the ocean." Climate was no longer primarily considered an issue of the field of geography, but of meteorology and oceanography, and climate science became "physics of climate" (e.g., Peixoto and Oort, 1992).

Since the 1970s, the notion that unconstrained emissions of greenhouse gases into the atmosphere generated by human activities will lead to significant changes of climatic conditions — a theory first proposed by Svante Arrhenius (1896) — was supported by evidence of a broad warming and finally embraced by the majority of climate scientists. The series of Assessment Reports by the *Intergovernmental Panel on Climate Change* (IPCC) is central to and documents this change. In the 1990s, human-driven climate change became the absolutely dominant topic in climate sciences (Weart, 1997, 2010). Climate research became to a large extent driven by concern with human-made climate and its impacts.

Unnoticed by most climate scientists, the developments in the last decades represent a return to the original but transformed anthropocentric view of the issue of climate (Stehr and von Storch, 2010). In contrast to the perspective of "climatic determinism," it was no longer the idea that climate determines the functioning and fate of societies, but that climate conditions human societies (Stehr and von Storch, 1997).

3.3 Methodological challenges of the physics of climate

In this section, we outline, after a brief retrospect of the success of physics, several concepts in climate science, which are not normally met in conventional physics — and thus represent serious obstacles from a physics point of view. One of the obstacles is the absence of "the equations" and the need for parameterizations; another is the difficulty to "predict" and finally the issue of inhomogeneity of data.

The pillars of the success story of physics in the last two centuries are the unbiased observation and description of natural phenomena, the reproducibility of experimental data, and the mathematical description of the empirical results leading to a generalization of the experimental results and the elucidation of the underlying basic laws of nature. Perhaps the most prominent example is the Newtonian classical mechanics which Newton developed on the basis of Kepler's observations and Galileo's gravity experiments. In classical mechanics, the time evolution of a system (like the motion of the earth around the sun) follows Newton's equations. The important thing is that, when the state of the system is known for a certain time (for the earth-sun system, this is the position and velocity of the earth relative to the sun), then the time evolution of the system can be calculated rigorously, and precise predictions can be made. By solving Newton's equations, one can predict, e.g., the trajectories of rockets, satellites, and space ships, which is the basis for modern space science.

Another example is electrodynamics which was established by Maxwell and based on the experimental and theoretical work by Coulomb, Volt, Ampere, Gauss and others. Like Newton's equations, the celebrated Maxwell equations describe comprehensively all (classical) electrical and magnetic phenomena, and not only those that they aimed to describe initially. Among others, Maxwell's theory led to the recognition that light is an electro-dynamical phenomenon.

Prerequisites of the success of physics were:

1. The departure from the anthropocentric view of life that for the first time allowed an unbiased view onto the natural phenomena (like planetary motion);
2. A new practice of publication: The protagonists did no longer (like the alchemists) hide their results but made them available to the public, allowing colleagues to reproduce (or falsify) them; and finally
3. The norm of checking theoretical hypotheses experimentally. In the case of conflicting theories, an *experimentum crucis* is needed to decide which theory is correct. Perhaps the most important *experimentum crucis* is the Michelson experiment on the velocity of light, which forms the basis for Einstein's theory of relativity.

In climate science, at least two of these requisites do not exist. Climate science has become anthropocentric and *experimenta crucis* are not possible. When approaching the subject of climate from a physics or mathematical point of view, the first question usually is — what is included, and how to describe it? What are its equations? The climate system has different "compartments," such as atmosphere, ocean, sea ice, land surface including river networks, glaciers and ice sheets, but also vegetation and cycles of substances, in particular, greenhouse gases (Fig. 1). An important element of the dynamics is given by fluid dynamics of the atmosphere, ocean and ice, which are described by simplified Navier–Stokes equations. However, due to the unavoidable discrete description of the system, turbulence cannot be described in mathematical accuracy, and the equations need to be "closed" — the effect of friction, in particular, at the boundaries between land, atmosphere and ocean, needs to be "parameterized" (e.g., Washington and Parkinson, 2005). Additional equations describing the flow and transformations of energy are needed — part of this may be described by the first law of thermodynamics, the conservation of energy. In these equations, we find source and sink terms, which are related to phase changes (condensation, for instance) and the interaction of cloud water and radiation. The sources and sinks often take place at the smallest scales and require additional state variables (such as the size spectrum

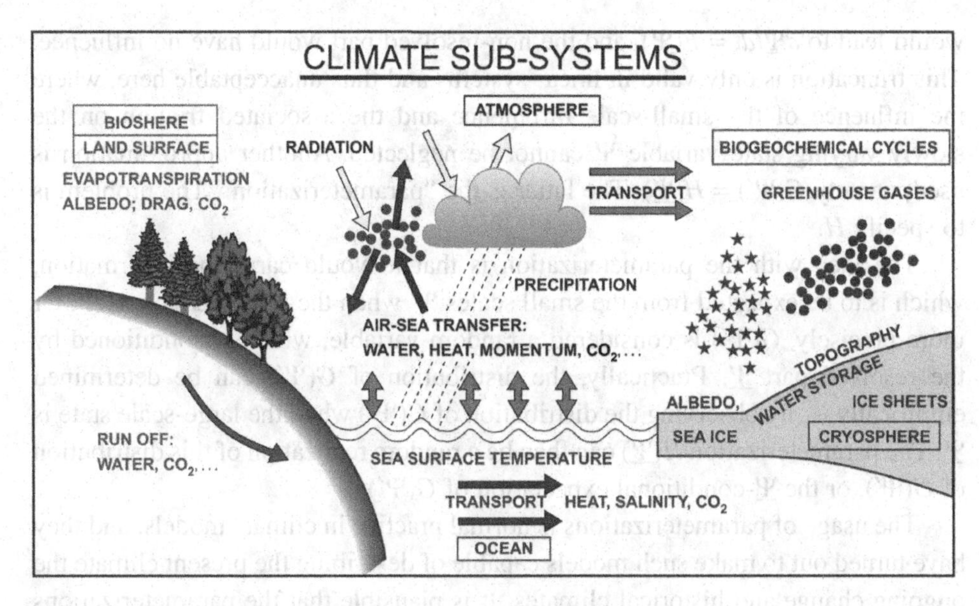

Figure 1: Schematic sketch of processes and variables in the climate system (Reprinted with permission of Klaus Hasselmann).

of cloud droplets). Again, such processes cannot be taken care of explicitly — and need to be "parameterized."

This issue of parameterization is difficult to understand (Müller and von Storch, 2004). The basic idea is that there is a set of "state variables" Ψ (among them, the temperature field at a certain time t at certain discrete positions on the globe), which describe the system, and which dynamics is given by a differential equation $d\Psi/dt = F(\Psi)$. The function F is nonlinear in Ψ and only approximately known. A rigorous analytical or numerical solution of the equations (as for Newton's equations) is impossible.

To simplify the equations and make them tractable for a numerical treatment, one splits each of the state variables C into a slowly and a rapidly varying component, $\Psi = \underline{\Psi} + \Psi'$.[1] $\underline{\Psi}$ represents that part of C, which is well represented with the given spatial resolution (say 100 km), and Ψ' being the unresolved part of smaller spatial scale. The equations are then approximately written as $d\underline{\Psi}/dt = F(\underline{\Psi}) + G(\Psi')$. Here, $F(\underline{\Psi})$ describes the influence of the resolved part $\underline{\Psi}$ on the future development of Ψ, whereas $G(\Psi')$ describes the influence of the non-resolved part, which is, of course, unknown. A conventional truncation of the equations

[1] In certain cases, this can be done by expanding Ψ into a Fourier series of trigonometric functions or spherical harmonics. Those with small wavenumbers then make up $\underline{\Psi}$ and the rest Ψ'.

would lead to $d\underline{\Psi}/dt = F(\underline{\Psi})$, and the non-resolved part would have no influence. This truncation is only valid in linear systems and thus unacceptable here, where the influence of the small-scale turbulence and the associated friction on the slowly varying state variable $\underline{\Psi}$ cannot be neglected. Another approximation is used, namely $G(\Psi') = H(\underline{\Psi})$. The latter is the "parameterization." The problem is to specify H.

The idea with the parameterization is that it would carry the information, which is to be expected from the small scales Ψ' when the resolved state is $\underline{\Psi}$. Or more precisely, $G(\Psi')$ is considered a random variable, which is conditioned by the resolved part $\underline{\Psi}$. Practically, the distribution of $G(\Psi')$ can be determined empirically — by observing the distribution of $G(\Psi')$ when the large-scale state is $\underline{\Psi}$. The parameterization $H(\underline{\Psi})$ can then be a random realization of this distribution of $G(\Psi')$, or the $\underline{\Psi}$-conditional expectation of $G(\Psi')$.

The usage of parameterizations is normal practice in climate models, and they have turned out to make such models capable of describing the present climate the ongoing change and historical climates. It is plausible that the parameterizations are valid "closures" also in a different climate [after all, in terms of physical (but not societal) magnitudes, any climate change would represent only a minor change], but the final evidence for this belief will be available only after the expected changes have taken place, have been observed and analyzed.

There are two important aspects of parameterizations.

One is a linguistic aspect, namely that in the language of climate modelers, parameterizations are named "physics," a shorthand for "unresolved physical processes." For a person uncommon with the culture of climate sciences, this terminology may go with the false connotation that parameterizations would be derived from physical principles. While the functional form of the parameterization $H(\underline{\Psi})$ may be motivated by a physical plausibility argument, the specific parameters used are either guessed, fitted to campaign or laboratory data, or to make the model skillful in reproducing the large-scale climate $\underline{\Psi}$. Thus, the word "physics" points to semiempirical "tricks."

Another aspect of parameterizations is their strong dependence on the spatial resolution. When the model is changed to run on a higher resolution, the parameterizations need to be reformulated or respecified. There is no rule how to do that when the spatial resolution is increased — which means that the difference equations do not converge towards a pre-specified set of differential equations, or, in other words: There is nothing like a set of differential equations describing the climate system per se, as is the case in most physical disciplines.

To summarize: In climate science, there are not "the equations" but only useful approximations, which crucially depend on the spatial resolution of the system.

This aspect causes many misunderstandings, in particular among mathematicians and physicists who often enough demand to see "the equations."

Unlike most physics disciplines, climate cannot be associated with spatial and temporal scales in a certain limited range — instead, climate varies on all spatial scales (on Earth) and extends across several magnitudes of timescales, from short-term events, measured in seconds, via timescales of decades and centuries, to geological timescales of millennia and more years. If we look at the relevant processes in the climate components atmosphere and ocean, we find a continuum of scales, as displayed in Fig. 2. The implication is that there are hardly independent observations from different locations, and the temporal memory extends across many decades of years.

As a practical rule, the World Meteorological Organization (WMO) mandated a hundred or more years ago that 30-year time intervals would represent "normal climatic conditions;" every 30 years new normals are determined. If we accept this somewhat arbitrary number of 30 years, we have to wait about 30 years to get a new realization of the climate system, which is at least somewhat independent of

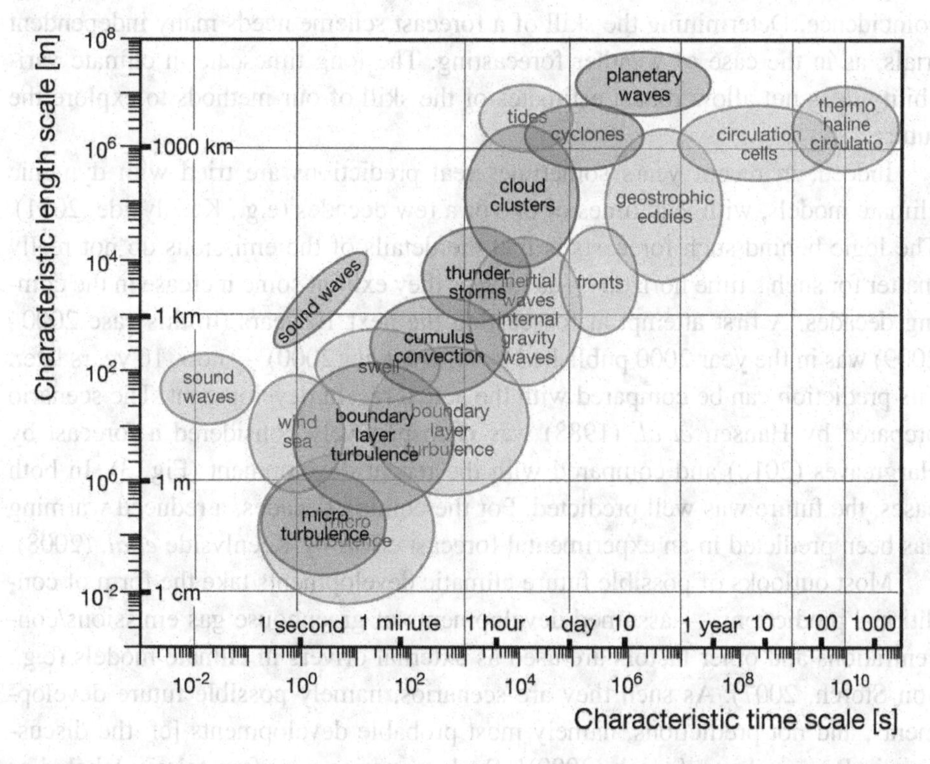

Figure 2: Spatial and temporal scales of processes in the climate components atmosphere and ocean.

previous states. Thus, tests of hypotheses, derived from historical data, using new data are hardly possible.

Real experiments, in the sense of paired configurations, which differ in a limited number of known details, are, of course, also not possible in the real world (as in any other geophysical setup). However, with quasi-realistic models, which serve as a kind of virtual laboratory, it is possible to perform virtual experiments, for instance, on the effect of different formulations of clouds and different specifications of physiographic detail, but also on elevated greenhouse gas concentrations in the atmosphere. Independent realizations can be generated; extended long simulations are possible so that the weather noise may be reduced and the looked-after signals, caused by an imposed experimental change in the system, may be more easily isolated. The problem is, of course, that even if the models share indeed many properties with reality, it is unproven if the specific model response is realistic.

Real forecasts are also hardly possible: Even if we are able to prepare a successful forecast for the coming 10 or 30 years, we cannot claim the "success" of our prediction scheme, because a single success may also have taken place by coincidence. Determining the skill of a forecast scheme needs many independent trials, as in the case of weather forecasting. The long timescale in climate variability does not allow robust estimates of the skill of our methods to explore the future.

Indeed, in recent years sometimes real predictions are tried with dynamic climate models, with lead times of one or a few decades (e.g., Keenlyside, 2011). The logic behind such forecasts is that the details of the emissions do not really matter for such a time horizon — as long as they exhibit some increase in the coming decades. A first attempt at forecasting the next 10 years (in this case 2000–2009) was in the year 2000 published by Allen *et al.* (2000) — now, 10 years later, this prediction can be compared with the actual recent development. The scenario prepared by Hansen *et al.* (1988) was retrospectively considered a forecast by Hargreaves (2010) and compared with the recent development (Fig. 3). In both cases, the future was well predicted. For the coming decades, a reduced warming has been predicted in an experimental forecast effort by Keenlyside *et al.* (2008).

Most outlooks of possible future climatic developments take the form of conditional predictions — assumed developments of greenhouse gas emissions/concentrations and other factors are used as external drivers in climate models (e.g., von Storch, 2007). As such they are scenarios, namely possible future developments, and not predictions, namely most probable developments [cf. the discussion in Bray and von Storch (2009)]. Such scenarios are often falsely labeled as "predictions" in the media, and even by some research institutions. They are

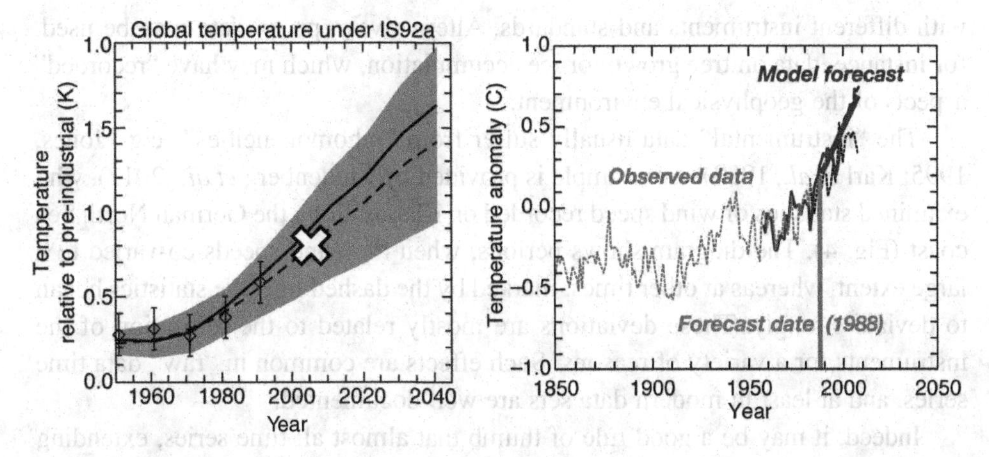

Figure 3: Left: Allen *et al.*'s (2000) forecast of global temperature made in 1999. Solid line shows original model projection. Dashed line shows prediction after reconciling climate model simulations with the HadCRUT temperature record, using data to August 1996. Grey band shows 5–95% uncertainty interval. White cross shows observed decadal mean surface temperature for the period 1 January 2000 to 31 December 2009 referenced to the same baseline. Right: Hansen's scenario published in 1988 as a prediction up to 2010 (redrawn after Hargreaves, 2010).

prepared with quasi-realistic climate models (e.g., Müller and von Storch, 2004), often abbreviated by GCM (which historically stands for General Circulation Models and not for Global Climate Models).

To summarize: Most future outlooks available to the scientific community and presented to the general public are not descriptions of most probable futures (predictions) but plausibly consistent and possible futures (scenarios or projections). In a few cases, real predictions have been prepared for the nearer future, and they have turned out to point into the right direction.

While this is encouraging, such sporadic successes cannot be considered as significant evidence for the general validity of climate models. At the same time, evidence is not available that would positively disqualify such models for being valid tools to study man-made climate change.

The lack of the option to do experiments prevents many uncertainties from being resolved, as, for instance, the climate sensitivity (temperature increase after equilibration when CO_2 concentrations are doubled). Indirect evidence is used for improving the estimate of such uncertain quantities, but some uncertainty remains. This leaves a certain range for interpreting the policy implications differently.

Because of the long waiting time for getting a new realization of the climate system, climate science must rely on historical "instrumental" data, data which have been measured for often quite different purposes, under different conditions, with different instruments and standards. Alternatively, proxy data may be used, for instance, data on tree growth or ice accumulation, which may have "recorded" aspects of the geophysical environment.

The "instrumental" data usually suffer from "inhomogeneities" (e.g., Jones, 1995; Karl *et al.*, 1993). An example is provided by Lindenberg *et al.* (2012), who examined statistics of wind speed recorded on islands along the German North Sea coast (Fig. 4). The diagram shows periods, when the wind speeds co-varied to a large extent, whereas at other times, marked by the dashed line, the statistics began to deviate strongly. These deviations are mostly related to the relocation of the instruments for a variety of reasons. Such effects are common in "raw" data time series, and at least in modern data sets are well documented.

Indeed, it may be a good rule of thumb that almost all time series, extending across several decades of years, suffer from some inhomogeneities — the more easily detectable inhomogeneities are "abrupt," such as those in Fig. 4, but the more difficult to detect are continuous changes. An example is the effect of continuous urbanization, which can be separated from the natural variability only within large error bars (Lennartz and Bunde, 2009).

Before using such data in climate analysis, the series have to be "homogenized" (e.g., Peterson *et al.*, 1998). For scientists and lay people, with insufficient insight into the contingencies of climate data, this significant hurdle is hardly

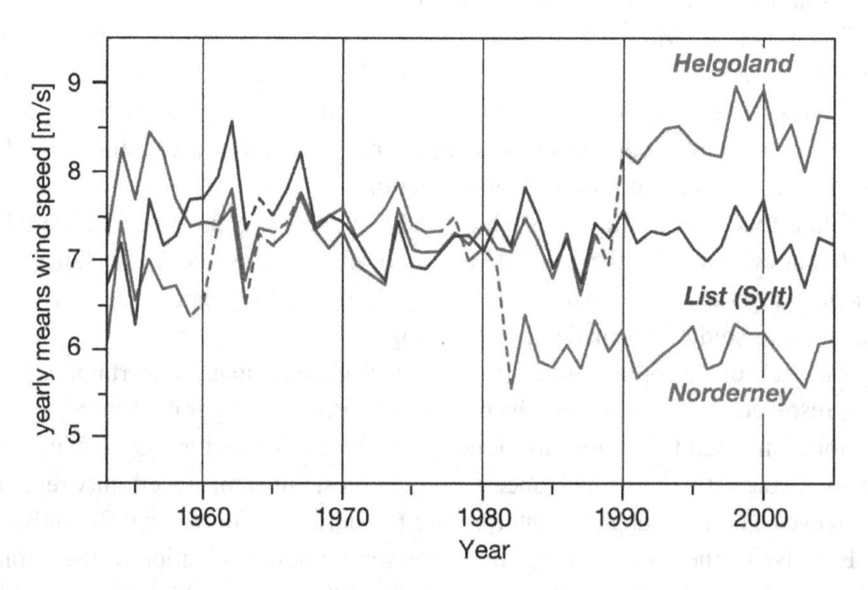

Figure 4: Yearly means of wind speed measurements from three synoptic island stations. Dashed lines label years with known station relocations (after Lindenberg *et al.*, 2012).

recognized. Therefore, it happens every now and then that publications show surprising results, which in the end display changing data recording practices and not changes in the climate system. A nice example is the conjectured increase of the absolute number of deep cyclones in the last century, which is due to insufficient data knowledge (Schinke, 1992). Also, contributors on weblogs often ask for "raw data," with the implicit suspicion that somebody may have tampered with the raw data in order to obtain preconceived results. This is in most cases not a wise approach — because of these invisible inhomogeneities (cf. Böhm, 2010).

"Proxy" data have other problems (Briffa, 1995). The main problem is that the proxies, for instance, growth rings in trees, or annual layers of sedimentary material, record not only some climate parameters, such as summer temperature but also other influences. The fundamental problem is that only part of the variability in the proxies is related to climate variability, in particular temperature. The proportion of variability, which may be related to climate drivers, differs in time, and the empirically derived transfer functions may show different amplitudes for different timescales. The famous problem of the hockey stick-named temperature reconstruction, which was based mostly on tree rings, had much to do with the nonuniform representation of long-term and short-term variability. An interesting exchange about proxy methodologies and robust claims-making is provided by a series of papers, comments and replies by Christiansen *et al.* (2009), Rutherford *et al.* (2010) and Christiansen *et al.* (2010).

To summarize: "data" entail complex issues in climate research. Historically collected "instrumental" data often suffer from inhomogeneities, related to changing observational, archival and analytical practices. Indirect proxy data provide information about changing physical conditions but compete with other unknown influences — so that stationarity and timescale dependence of the information content of such data are an issue. More expertise about the process of using instruments and of storing influences in indirect data is required.

Thus, there are series of obstacles and uncertainties, which are uncommon in conventional physics, and represent special challenges when dealing with climate dynamics and impacts.

3.4 The sociocultural context

There is another set of factors that makes the science of climate "different" from other natural science fields — namely that climate research, the issues, results and individuals — are firmly embedded into sociohistorical, sociocultural and socioeconomic contexts. This is already illustrated by virtue of the fact that most, perhaps almost all, of present climate research activities are related to the issue of anthropogenic climate change and its impact on the natural environment and society.

The main issue in the societal context concerns the statistics of weather (in atmosphere and ocean) and its changes, such as the frequency and intensity of extreme events such as storms, heatwaves and flooding. Weather statistics are significant data for societies, their infrastructure and inhabitants because they contain important information about possible impacts (and adaptive measures to deal with them) and options for keeping a check on the drivers (mitigation). Both strategies in response to a changing climate, that is, reducing emissions and reducing vulnerability, are subjects of a wide range of scientific fields including the engineering sciences, hydrology, law, geography, policy sciences, ecology, economy and social sciences.

Thus, climate research has significant attributes beyond physics. We could now start to discuss the needed contributions from these other fields, but we do not. We instead concentrate more on the functioning of the science–society knowledge interaction.

This interaction of climate research with society in general and with policy making in particular is linked to the joint presence of two factors. One factor that we have just discussed above is the high uncertainty about the "facts" of climate dynamics, ranging from the climate sensitivity to regional specification, to the presence of other social drivers and to future options of dealing with emissions and impacts. The other factor concerns the societal response to climatic conditions, how we interpret and deal with climate-related processes. Our ordinary everyday understanding of climate is closely related to our way of life, mediated, of course, by the way in which the mass media shape climate issues according to media logic (Weingart *et al.*, 2000; Boykoff and Timmons, 2007; Carvalho, 2010). References to climate and climate change in public communication may be employed, for example as a tool to legitimate changes to our way of life, or, in the opposite sense, as a means to defend dominant world views.

Under these circumstances, but not only because the nature of our understanding of climate is embedded in everyday life, climate science becomes along with other modern scientific fields "postnormal" (Funtovicz and Ravetz, 1985; Bray and von Storch, 1999). A broad range of essentially contested terms and explanations enter the public arena and compete for attention, accounts that may also be brought in position to give credence to different world views and legitimacy to political and economic interests.

There seem to be two major contending classes of explanations of the climate and climate change (von Storch, 2009). One, which we label as "scientific construct" of human-made climate change, states that processes of human origin are influencing the climate — that human beings are changing the global climate. In almost all localities, at present and in the foreseeable future, the frequency distributions of the temperature continue to shift to higher values; sea level is rising; and amounts of rainfall are changing. Some extremes such as heavy rainfall events will change. The driving force behind alterations beyond the range of natural variability is above all the emission of greenhouse gases, in particular CO_2 and methane, into the atmosphere, where they interfere with the radiative balance of the Earth system.

The scientific construct is widely supported within the relevant scientific communities and has been comprehensively formulated particularly, thanks to the collective and consensual efforts of the UNO climate council, the IPCC.[2] Of course, there is not a complete consensus on all aspects of the construct in the scientific community so that speaking of "the scientific construct" is somewhat of a simplification. What is consensual and enumerated in the previous paragraph is the core of the scientific construct.

A different conception of climate and climate change may be labeled the social or cultural construct (cf. Stehr and von Storch, 2010). In the context of this concept, climate and weather patterns are also changing, the weather is less reliable than it was before, the seasons less regular and the storms more violent. Weather extremes are taking on catastrophic and previously unknown forms.

What causes these changes in weather patterns? A variety of economic reasons and psychological motives tend to be adduced, for example, sheer human greed and simple stupidity. The mechanism that is at work may be described as follows: Nature is retaliating and striking back. For large segments of the population, at least in central and northern Europe, this mechanism producing climate change is taken for granted. In older times, and even sometimes today, adverse weather patterns were the prompt response of the gods angered by human sins (e.g., Stehr and von Storch, 2010).

The cultural construct of climate and changing weather patterns takes many different forms depending on the traditions in a society, its development and dominant aspirations — but what is described above as the everyday concept of climate and weather represents something like a standard core of such statements.

---

[2]The support among climate scientists seems indeed very broad, when related to the key assertions just listed (cf. Bray and von Storch, 2007). Whether the emergence of errors in Working Group (WG) 2, and possibly WG3, of 4th Assessment Report (AR4) of IPCC, which so far all point towards a dramatization, and after it became known that a key data set (CRU) could no longer be reproduced because of some original data having been "lost," will have implications for this support within climate science and the general scientific community remains to be seen.

Obviously, the scientific construct is hardly consistent with such cultural constructs.

The position of so-called "climate skeptics" is not discussed here because there is no consistent body of knowledge of "skeptic" climate science but merely a collection of various, often highly contested issues that range from detailed matters to much more general assertions, e.g., that greenhouse gases would have no significant impact on climate. The absence of a consistent body of assertions does not imply in principle that the questions raised by 'skeptics' might not in one or the other instance be helpful to constructively move the science of climate forward.

In this postnormal situation where science cannot make concrete statements with high certainty, and in which the evidence of science is of considerable practical significance for formulating policies and decisions, this science is impelled less and less by the pure "curiosity" that idealistic views glorify as the innermost driving force of science, and increasingly by the usefulness of the possible evidence for just such formulations of decisions and policy (Pielke, 2007b). It is no longer being scientific that is of central importance, nor the methodical quality, nor Popper's dictum of falsification, nor Fleck's (1980) idea of repairing outmoded systems of explanation; instead, it is social and political utility of knowledge claims that carry the day. Not correctness, nor objective falsifiability, occupies the foreground, but rather social acceptance and social utility.

In its postnormal phase, science thus lives on its claims, on its staging in the media and on its affinity and congruity with sociocultural constructions. These knowledge claims are not only raised by established scientists, but also by other, self-appointed experts, who often are bound to special interests. Representatives of social interests seek out those knowledge claims that best support their own position. One need to only recall the Stern report [see the critique by Pielke (2007a) or Yohe and Tol (2008)] or the press releases of US Senator Inhofe.

3.5 Conclusions

Two major conclusions about the science of climate and the knowledge about climate may be drawn.

The scientific construct is mostly based on a physical analysis of climate and developed by natural scientists. It describes the left two blocks in Fig. 5. In the "linear model" (Beck, 2010; Hasselmann, 1990), the middle blocks, representing social and cultural dynamics, are not taken into account. Instead, once society has given a metric of determining "good" and "less good," it is simply a matter of understanding the "physical" (including economic) system.

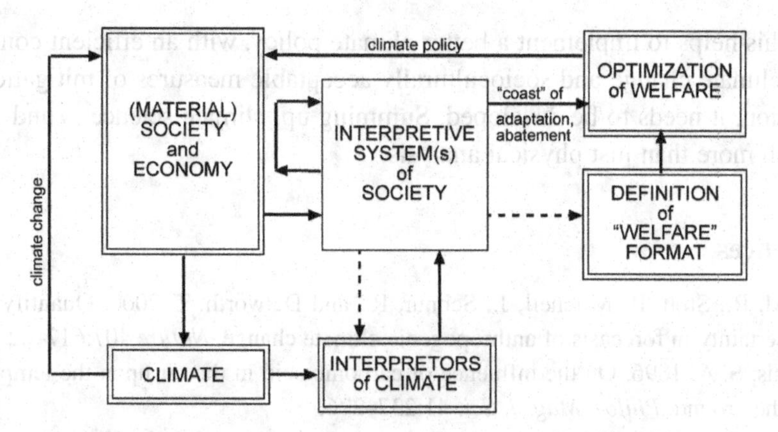

Figure 5: The perceived climate and society model (after Stehr and von Storch, 2010).

However, the climate scientists are also part of society and not immune to dominant societal conceptions of the nature and the impact of climate and climate change on human conduct, they tend to embed their analysis, especially in efforts to communicate their knowledge to policy makers and society at large in ways which are attentive to the sociocultural construct of climate and climate change. It is not surprising that in this postnormal situation scientists concerned about the impact of the greenhouse gases, in their desire to save the world, may develop some bias towards an overdramatization. The discussion itself often resembles more a religious than a physics discussion where the nonbelievers (of the role of the greenhouse gases and their impact) are called "deniers."

One therefore is able to surmise that the transfer of the scientific construct into the societal realm goes along with a subtle transformation of the climate knowledge, by blending the scientific construct with the sociocultural construct (the middle blocks in Fig. 5). Obviously, in the model described in Fig. 5, the basic assumption of physics, that there are given quantifiable laws (linear or nonlinear), is no longer valid. Understanding the interaction of climate and society is not only an issue of physical analysis (with laws) but of society/culture analysis (without laws) as well.

Obviously, the situation is not quite that straightforward, it is not easily deconstructed, and the interrelations of scientific and everyday construct are difficult to dissemble. To comprehend and disentangle the multiple interactions of science and society in the case of our understanding of climate and climate change is nonetheless a real and worthy scientific and practical challenge. It needs a trans-disciplinary approach, bringing together scientists with a solid background in physics, and scholars who understand societal and knowledge dynamics (Pielke, 2007b).

If this helps to implement a better climate policy, with an efficient constraining of climate change and socioculturally acceptable measures of mitigation and adaptation, it needs to be developed. Summing up, climate science is and should be much more than just physical analysis.

References

Allen, M. R., Stott, P., Mitchell, J., Schnur, R., and Delworth, T. 2000. Quantifying the uncertainty in forecasts of anthropogenic climate change. *Nature* 407:617–620.

Arrhenius, S. A. 1896. On the influence of carbonic acid in the air upon the temperature of the ground. *Philos. Mag. J. Sci.* 41:237–276.

Arrhenius, S. A. 1908. *Das Werden der Welten*. Leipzig: Akademische Verlagsanstalt.

Beck, S. 2010. Moving beyond the linear model of expertise? IPCC and the test of adaptation. *Reg. Environ. Change*. doi:10.1007/s10113–010–0136–2.

Böhm, R. 2010. 'Faking versus adjusting' — why it is wise to sometimes hide 'original' data. <http://klimazwiebel.blogspot.com/2010/01/guest-contribution-from-reinhard-bohm.html> (as of 22 March 2010).

Boykoff, M. T. and Timmons, R. J. 2007. Media Coverage of Climate Change: Current Trends, Strengths, Weaknesses. Human Development Report 2007/2008.

Bray, D. and von Storch, H. 1999. Climate science: An empirical example of postnormal science. *Bull. Am. Meteorol. Soc.* 80:439–456.

—— 2007. Climate Scientists' Perceptions of Climate Change Science. GKSS-Report 11/2007.

—— 2009. 'Prediction' or 'projection'? The nomenclature of climate science. *Sci. Comm.* 30:534–43. doi:10.1177/1075547009333698.

Briffa, K. R. 1995. Interpreting high-resolution proxy climate data — the example of dendro-climatology. In H. von Storch and A. Navarra (eds.), *Analysis of Climate Variability: Applications of Statistical Techniques*. Berlin: Springer Verlag, pp. 77–84.

Carvalho, A. 2010. Media(ted) discourses and climate change: A focus on political subjectivity and (dis)engagement. *Clim. Change* 1(2). doi:10.1002/wcc.13.

Christiansen, B., Schmith, T., and Thejll, P. 2009. A surrogate ensemble study of climate reconstruction methods: Stochasticity and robustness. *J. Clim.* 22:951–976.

—— 2010. Reply to comment on 'A surrogate mensemble study … ' by Rutherford *et al.* *J. Clim.* 23:2839–2844.

Curry, J. A. and Webster, P. J. 2011. Climate science and the uncertainty monster. *Bull. Am. Meteorol. Soc.* 92:1667–82. https://doi.org/10.1175/2011BAMS3139.1.

Fleck, L. [1935] 1980. *Entstehung und Entwicklung einer wissenschaftlichen Tatsache: Einführung in die Lehre vom Denkstil und Denkkollektiv.* Frankfurt am Main: Suhrkamp.

Fleming, J. R. 1998. *Historical Perspectives on Climate Change*. Oxford: Oxford University Press.

Friedman, R. M. 1989. *Appropriating the Weather: Vilhelm Bjerknes and the Construction of a Modern Meteorology.* Ithaca, NY: Cornell University Press.

Funtowicz, S. O. and Ravetz, J. R. 1985. Three types of risk assessment: A methodological analysis. In C. Whipple and V. T. Covello (eds), *Risk Analysis in the Private Sector*. New York: Plenum, pp. 217–31.

Hann, J. 1903. *Handbook of Climatology. Vol. 1: General Climatology*. New York: Macmillan.

Hansen, J., Fung, I., Lacis, A., Rind, D., Lebedeff, S., Ruedy, R., Russell, G., and Stone, P. 1988. Global climate changes as forecast by Goddard Institute for Space Studies three-dimensional model. *J. Geophys. Res. — Atmos.* 93(D8):9341–9364.

Hargreaves, J. 2010. Skill and uncertainty in climate models. *Wiley Interdiscip. Rev. Clim. Change* 1:556–64. doi:10.1002/wcc.58.

Hasselmann, K. 1990. How well can we predict the climate crisis? In H. Siebert (ed.), *Environmental Scarcity: The International Dimension*. Tübingen: JCB Mohr.

Jones, P. D. 1995. The instrumental data record: Its accuracy and use in attempts to identify the 'CO_2 Signal'. In H. von Storch and A. Navarra (eds.), *Analysis of Climate Variability: Applications of Statistical Techniques*. Berlin: Springer Verlag.

Karl, T. R., Quayle, R. G., and Groisman, P. Y. 1993. Detecting climate variations and change: New challenges for observing and data management systems. *J. Clim.* 6:1481–1494.

Keenlyside, N. S. 2011. Prospects for decadal climate prediction. *Wiley Interdiscip. Rev. Clim. Change* 1(5):627–35. doi:10.1002/wcc.69.

—— Latif, M., Jungclaus, J., Kornblueh, L., and Roeckner, E. 2008. Advancing decadal-scale climate prediction in the North Atlantic sector. *Nature* 453: 84–88.

Lennartz, S. and Bunde, A. 2009. Trend evaluation in records with long-term memory: Application to global warming. *Geophys. Res. Lett.* 36:L16706.

Lindenberg, J., Mengelkamp, H.-T., and Rosenhagen, G. 2010. Representativity of near surface wind measurements from coastal stations at the German Bight. *Meteorol. Z.* 21:99–106. doi:10.1127/0941-2948/2012/0131.

Müller, P. and von Storch, H. 2004. *Computer Modelling in Atmospheric and Oceanic Sciences — Building Knowledge.* Berlin: Springer Verlag.

Peixoto, J. P. and Oort, A. H. 1992. *Physics of Climate*. American Institute of Physics.

Peterson, T. C., Easterling, D. R., Karl, T. R., Groisman, P., Nicholls, N., Plummer, N., Torok, S., Auer, I., Boehm, R., Gullett, D., Vincent, L., Heino, R., Tuomenvirta, H., Mestre, O., Szentimrey, T., Saliner, J., Førland, E., Hanssen-Bauer, I., Alexandersson, H., Jones, P., and Parker, D. 1998. Homogeneity adjustments of *in situ* atmospheric climate data: A review. *Intern. J. Climatol.* 18:1493–1517.

Pielke, R. A., Jr. 2007a. Mistreatment of the economic impacts of extreme events in the Stern Review Report on the economics of climate change. *Glob. Environ. Change* 17:302–310.

—— 2007b. *The Honest Broker*. Cambridge: Cambridge University Press.

Rutherford, S. D., Mann, M. E., Ammann, C. M., and Wahl, E. R. 2010. Comment on: 'A surrogate ensemble study of climate reconstruction methods: Stochasticity and robustness' by Christiansen, Schmith, and Thejll. *J. Clim.* 23:2832–2838.

Schinke, H. 1992. Zum Auftreten von Zyklonen mit niedrigen Kerndrücken im atlantisch-europäischen Raum von 1930 bis 1991. *Wiss. Z. Humboldt Univ. Berlin, R. Math. Naturwiss.* 41:17–28.

Solomon, S., Qin, D., Manning, M., Marquis, M., Averyt, K., Tignor, M. M. B., Le Roy Miller, H. Jr, and Chen, Z. (eds.) 2007. *Climate Change 2007: The Physical Basis.* Cambridge: Cambridge University Press.

Stehr, N. and von Storch, H. 1997. Rückkehr des Klimadeterminismus? *Merkur* 51:560–562.

——1999. An anatomy of climate determinism. In H. Kaupen-Haas (ed.), *Wissenschaftlicher Rassismus — Analysen einer Kontinuität in den Human- und Naturwissenschaften.* Frankfurt. a. M.: Campus-Verlag, pp. 137–185.

—— 2010. *Climate and Society: Climate as a Resource, Climate as a Risk.* Singapore: World Scientific.

von Humboldt, A. von [1845] 1864. *Cosmos: Sketch of a Physical Description of the Universe.* Vol. i. London: Henry G. Bohn.

von Storch, H. 2007. Climate change scenarios — purpose and construction. In H. von Storch, R. S. J. Tol, and G. Flöser (eds.), *Environmental Crises: Science and Policy.* Heidelberg: Springer, pp. 5–15. ISBN 978-3-540-75895-2, 5-1.

—— 2009. Climate research and policy advice: Scientific and cultural constructions of knowledge. *Environ. Sci. Pol.* 12:741–747.

—— and Stehr, N. 2000. Climate change in perspective: Our concerns about global warming have an age-old resonance. *Nature* 405:615.

Washington, W. M. and Parkinson, C. L. 2005. *An Introduction to Three-Dimensional Climate Modelling*, 2nd edn. Sausalito, CA: University Science Books.

Weart, S. R. 2011. The development of the concept of dangerous Anthropogenic Climate Change. In J. S. Dyzek *et al.* (eds.), *The Oxford Handbook of Climate Change and Society*. New York: Oxford University Press, pp. 67–81.

—— 1997. The discovery of the risk of global warming. *Phys. Today* (January) 50:35–40.

—— 2010. The idea of anthropogenic global climate change in the 20th century. *Wiley Interdiscip. Rev. Clim. Change*. Volume 1. Published online: 22 December 2009. doi:10.1002/wcc.6.

Chapter 5

Climate Policies

While the work of Stehr and von Storch in the first half of their cooperation was mainly analytical, focusing on the competition between socially and scientifically constructed knowledge, on the different qualities of social and natural sciences in informing public opinion and advising policy discussions, the history of reflections about the relationship between societies and climate, the focus of the two authors subsequently expanded to include the political consequences of this knowledge about climate. This effort culminated in the "Zeppelin Manifesto" (see Section 1 in Chapter 6).

In the following section of our anthology, a total of six articles are reprinted. In Section 1, we discuss the need not only to consider the global problem of climate change but also the "the bottleneck of national, regional and even communal contingencies." The role of the communication is discussed in Section 2, and in Section 3, the question, how climate science can support in a sustainable manner the formation of policymaking. Claims voiced in parts of the scientific community and in public discourse that democratic governance is incapable of coping with a complex and long-term problem like climate change and may therefore require a "technocratic" fix are critically examined in Section 4.

When examining the public and political discourse in developed societies, it becomes obvious that the policy option of *adaptation* to climate change is widely seen as inadequate or even defeatist (Section 5). The political goal tends to be framed as "climate protection" — but not in the sense of protecting the population from the hazardous effects of climate change, but protection of climate itself from human activity (Section 6). The duality and convergence of the two policy options to confront the consequences of anthropogenic climate change, adaptation and mitigation are discussed in Section 7.

In addition to such scholarly articles, Nico Stehr and Hans von Storch presented their ideas also into the public arena, via newspapers, magazines, radio, television and social media.[1]

[1] For example, Stehr, N., und H. von Storch, 2019: Wir müssen lernen, uns anzupassen. *CICERO Online*, 30 October 2019.

von Storch, H., und N. Stehr, 2019: Die unsichtbaren Elefanten in der Klimapolitik. Iablis. https://www.iablis.de/iablis/themen/2019-formen-des-politischen/thema-2019/562-die-unsichtbaren-elefanten-in-der-klimapolitik.

Stehr, N., und H. von Storch, 2011: Forscher als Politikberater: Der Welt rettende Professor ist gescheitert. *Spiegel Online*, 11 December 2011.

von Storch, H., und N. Stehr, 2010: Thoughts on climate research and policy. *Newsletter of Europäischen Akademie Bad Neuenahr-Ahrweiler GmbH* 99, S. 1–2.

von Storch, H., und N. Stehr, 2010: Klimaforschung und Klimapolitik – Rollenverteilung und Nach haltigkeit. *Naturwissenschaftliche Rundschau* 63, nr. 744, 301–307.

Stehr, N., und H. von Storch, 2009: Die lästige Demokratie. *Spiegel Online*, 29 December 2009.

Stehr, N., und H. von Storch, 2009: Klimaschutz und Vorsorge, *forum* 291, 21–24.

von Storch, H., und N. Stehr, 2007: Anpassung an den Klimawandel. Aus Politik und Zeitgeschichte 47/2007, 19 November 2007 (English version: http://www.hvonstorch.de/klima/pdf/Parlament.english.pdf).

von Storch, H., und N. Stehr, 2007: Essay: Nur keine unbequeme Wahrheit. *ZEIT Wissen*, 18 October 2007.

von Storch, H., und N. Stehr, 2007: Der öffentliche Diskurs über das Klima oder – die politische Macht der Klimaforschung. In: Philipp Mißfelder (Hrsg.): *Umdenken. Für eine nachhaltige Klimapolitik*, S. 14–21.

Stehr, N., und H. von Storch, 2007: Einsichten in das Machbare. Klimaschutz ist vor allem auch Schutz vor dem Klima. *Süddeutsche Zeitung*, 15 September 2007.

Stehr, N., und H. von Storch, 2005: Trägheitsfaktor Natur. Anpassung statt Klimapolitik: Was New Orleans lehrt. *Frankfurter Allgemeine Zeitung*, 21 September 2005 – English version: The Sluggishness of Politics and Nature, http://www.hvonstorch.de/klima/pdf/NO.english.pdf.

von Storch, H., und N. Stehr, 2005: Klima inszenierter Angst [A Climate of Staged Angst]. *SPIEGEL 4/2005*: 160–161; http://www.hvonstorch.de/klima/pdf/SPIEGEL.prometheus.pdf.

1. Feasibility-Based Policies: Investigation Application*

Potential solutions to impending climate change must take into account a number of societal, national, and global challenges that are relevant today and will continue to be so in future. Among these challenges are the asymmetries in standards of living between the world's societies: The economic aspirations of North and South, the diverging positions of countries rich or poor in natural resources, of democratic and autocratic political regimes, and of states with dramatically different demographic dynamics. Last but not least, we also face the challenge of how to handle the strongly divergent convictions of what members of different cultures hold "sacred."

Considering the context of this situation as well as the unforeseeable events and developments that took place during the years following the Kyoto agreement, we have strong doubts that a consensual, globally effective strategy for the sustainable limitation of greenhouse gas emissions will be implemented any time soon. The failed political efforts of the past that aimed at sustainably protecting the global climate from the consequences of human actions are a sure indication that this skepticism is appropriate.

The implementation of global agreements must still pass through the bottleneck of national, regional, and even communal contingencies. There is no global political order in place that could support the implementation of global agreements or even enforce it by means of appropriate sanctions. Every political system will produce its own reactions to the challenges of climate change. The contradictoriness and fragility associated with any sort of (aggregated) activity are inevitable and will constitute the fundamental framework for any solution proposed in reaction to demands for timely and targeted action against climate change.

These elementary, and clearly contradictory, framework conditions for any sort of targeted action still feature insufficiently in the public climate debate and are often even treated as taboos.

What insights could therefore be brought to these political debates, misguided developments, and dead ends? What insights could promote an investigation of a feasibility-based policy and force the (often ideologically tainted) propagators of

*This section originally appeared in Stehr, N. and H. von Storch: "Feasibility-based policies: Investigation and application," pp. 14–21 in T.U. Graz (ed.), *Urbanity Not Energy. Urban Future Scenarios with Genuine Sustainability. Graz Architecture Magazine*. Volume 5. Vienna: Springer, 2008.

wishful thinking who loom large in the political circles of climate research to come back down to earth?

A few weeks ago, in about 2008, German chancellor Angela Merkel stated that carbon dioxide emissions must be limited to an average of two tons per person per year by the middle of this century to ensure that the disastrous consequences of climate change and wars over resources will be avoided. If not, the earth could heat up by more than the "critical threshold" of 2 degrees Celsius by 2050. As the average American citizen is currently producing twenty tons of carbon dioxide per year, the average German eleven tons per year, and a typical citizen of a developing country considerably less, her proposal can at least be interpreted as a tentative answer to the question of what would constitute a "fair" individual level of carbon dioxide pollution for a global population that is continuing to grow. For Germany, the target set by Merkel for 2050 would mean a reduction by 82 percent; for the U.S., by 90 percent.

The figures available on the average emission of carbon dioxide are controversial. For Germany, they most likely exceed the amount of eleven tons per citizen per year that is given here. Furthermore, the global values are at best indications of a carbon dioxide "justice quotient." But even if they are correct, they are inconsistent with realistic future expectations. In 2050, the population of the earth will have risen to 9 billion; today it is 6.5 billion. Carbon dioxide emissions of two tons per person would amount to total global emissions of 18 billion tons, a figure that would in any case be insufficient to stabilize the world's climate.

Meanwhile, actual carbon dioxide emissions continue to rise. Emissions are currently increasing, also in Germany. At the moment, we are moving more in the direction of fifteen instead of two tons per person per year.

Two other viewpoints are particularly relevant in this context: First, in an essay published in the *Geophysical Research Letters* earlier this year, H. Damon Matthews and Ken Caldeira (2008: L04705) come to the conclusion that a *stabilization* of the global temperature over the next few centuries is only possible if CO_2 emissions are reduced to zero: "In the absence of human intervention to actively remove CO_2 from the atmosphere (e.g., Keith et al., 2006), each unit of CO_2 emissions must be viewed as leading to quantifiable and essentially permanent climate change on centennial timescales. We emphasize that a stable global climate is not synonymous with stable radiative forcing, but rather requires decreasing greenhouse gas levels in the atmosphere. We have shown here that stable global temperatures within the next several centuries can be achieved if CO_2 emissions are reduced to nearly zero. This means that avoiding future human-induced climate warming may require policies that seek not only to decrease CO_2 emissions but to eliminate them entirely." It is no secret that this climate protection target will be difficult to reach; this makes preventive research and policies all the more pressing. The greater the success of mitigation, the better. In any case, the need for adaptive measures remains.

Secondly, in a study soon to be published, Peter Sheehan (2008) refers to new data on global economic growth and worldwide CO_2 emissions, pointing out that in recent years the world "has moved to a new path of rapid global growth, largely driven by the developing countries, which is energy intensive and heavily reliant on the use of coal — global coal use will rise by nearly 60 percent over the decade to 2010. It is likely that, without changes to the policies in place in 2006, global CO_2 emissions from fuel combustion would nearly double their 2000 level by 2020 and would continue to rise beyond 2030. Neither the SRES marker scenarios nor the reference cases assembled in recent studies using integrated assessment models capture this abrupt shift to rapid growth based on fossil fuels, centered in key Asian countries."[2] In short, the assumptions made so far on the future volume of global emissions are most likely too conservative. This means that efforts to reduce CO_2 emissions must be even more comprehensive in order to reach the desired goals of climate policy.

This in turn indicates that global efforts to limit greenhouse gas emissions will most likely be only moderately successful. Faced with these risks, technology optimists are beginning to consider large-scale technological possibilities to weaken the rate of climate change as alternatives to traditional "climate protection" based on reduction. Such possibilities include the mitigation of sun radiation or the depositing of CO_2 in the sea. At the moment, it hardly looks as if this option will receive the necessary political support, which means that a feasible climate policy will not only rely on research and political efforts but also, and to an increasing extent, on preventive and adaptive measures.

Why, then, are preventive strategies in climate policy and research, i.e., efforts to reduce the vulnerability of societies and their infrastructures with regards to the consequences of climate change, subjected to such extensive taboos, both by the media and by political players?

As early as fifteen years ago, former U.S. vice president Al Gore, now winner of an Oscar as well as the Nobel Peace Prize, voiced his uncompromising rejection of a climate policy based on adaptive strategies. Gore considers such a policy an expression of intellectual and political laziness, or, even worse, "an arrogant faith

[2] The IPCC Special Report on Emissions Scenarios (SRES; Nakicenovic *et al.*, 2000) discusses the qualitative societal framework conditions (e.g., political, social and cultural developments) that influence the emission volumes. The *SRES Emissions Scenarios* are the quantitative interpretations of this narrative.

in our ability to react in time to save our own skin." Only recently, Gore repeated this credo during a discussion of his film *An Inconvenient Truth* at Columbia University in New York. We must concentrate on reduction is Gore's uncompromising message to science, politics and society as a whole.

Al Gore's message is a fairly accurate echo of a climate-deterministic attitude that prevails both in science and in everyday life: Due to their unique power and influence over human life, natural forces — and the climate in particular — are responsible for a wide range of societal processes and regional particularities. Climate equals fate; it rules the successes and failures of entire civilizations. In other words, the influence of climate is inescapable. From this perspective, climate changes — whether man-made or "natural" — must by definition constitute an attack on the very foundations of any society.

Until recently, scientists and philosophers emphasized the sustainable effects of climate on the development of mankind. Although science has by now abandoned this sort of crude climate determinism, it still has not disappeared entirely from the current debate. When Gore and many other observers of climate change polemise against preventive measures, they in some ways fall victim to a school of thought that is considered outdated, an ideology.

In this school of thought, it is almost presumptuous to imagine that we could outsmart the world's climate — for instance, by means of technological tricks or preventive measures; strategies of this kind thus convey a wrong sense of security. Adapting ourselves to changing climate conditions thus represents the traditional human hubris in the face of natural forces. We believe that this philosophical assumption is behind the trivialization of adaptive and preventive strategies in the public debate of climate change on a scientific, political, and societal level. However, there are also other, equally significant reasons.

Let us begin with the reasons that can be attributed to the *scientific* investigation of climate change. Faced with recurring doubts, scientific efforts have so far concentrated on two topics: First of all, they aimed at proving that — seen within historical dimensions — we are currently experiencing a rapid and unique global climate change. Secondly, science focused on producing evidence that would show conclusively that the observed changes in the world's climate were caused by mankind itself. Climate science, in itself a young discipline, has reached these goals within the space of only a few years and has today achieved a far-reaching consensus, as reflected by the reports of the IPCC. Climate science has thus fulfilled one of its self-defined central functions, by showing that a man-made climate change is currently manifesting itself and that it will become more pronounced in the foreseeable future.

However, this consensus within climate science does not give rise to any indispensable, evidence-based instructions for countermeasures — much to the displeasure of science, but also of politics and its dominant vision of the instrumental efficacy of its findings. The dynamics of society are much more complex than those of climate. Nature's fluctuation time frames and time horizons simply do not correspond with the variety of phases and planning horizons in the lives of members of human societies. In comparison, the time frame of climate processes is sluggish and does not correspond with the possibilities and framework conditions of societal change, which are much more short-term in nature.

At this point, we must ask the question of what this climate change actually means, in a world that is already changing radically. This is a question that also concerns scientific disciplines beyond physically orientated climate research; for instance, climate impact scientists, in particular, social scientists, who must investigate how this global change, which involves much more than just climate change, may develop in future; and to what extent this development can be controlled or aided. So far, proposals to this effect have been mostly based on simple models drawn up by climate economists, attempting to reduce the problem to a small number of existential motives, but this approach is certainly too naïve. Our ideas of future societal conditions are shady at best, and the same is true for long-term technological and political frameworks. It is impossible to base definitive instructions for measures to be taken on these contours.

The societal status of natural science and technology is an important reason why the social sciences are reluctant to face the challenge of "climate change in a changing world." As long as the "human sciences" (Norbert Elias) will continue to occupy a subordinated position in society and their influence is systematically underestimated, the competence to solve the problem of climate change will continue to be primarily seen as the task of the natural sciences and of technology. One of the most frequent answers heard in this context is that, sooner or later, we must find radically new sources of energy. The question of how these efforts will be able to protect our existential foundations, now and in future decades, from the dangers of our future climate, which are severe already and will be even more pronounced in the future, is simply pushed aside.

The lack of prestige of the human sciences in society, in combination with the overconfidence of the natural and technological sciences, reduces the problem of climate change to a purely scientific-technological problem. On the market of public knowledge, the natural sciences are the first to offer their diagnosis but are then tied by the fact that their description of the situation requires a certain precise therapy. The road from insight to possible action is thus portrayed as clear-cut, as linear, as mandatory. It is not surprising that the terminology of medical science — the direct route from anamnesis to therapy — plays a central role in this context.

Another consequence of the special status enjoyed by the natural and technological sciences is that the failures of the therapy presented by publicly visible climate researchers as "mandatory," and the lack of resonance of any excursions into "alien" fields of research, are denounced as a regrettable backwardness of the collective mind, or as egotism on the part of politics and society. At times it almost seems as if, when this mandatory advice is not accepted, this egotism is "treated" by escalating the assumed potential dangers.

What we need is to rethink the preferential status of the natural and technological sciences, to work towards a social climate science that focuses on societal issues, and to gain political insights into feasibility-based solutions. In this context, what is feasible is a certain limitation of the emission of greenhouse gases that are damaging to the climate, but primarily also a protection of society from a climate that changes rapidly.

Bibliography

D. W. Keith, M. Ha-Duong, J. K. Stolaroff, "Climate strategy with CO_2 capture from the air," *Climatic Change*, 74/2006, 17–45.

H. D. Matthews, K. Caldeira, "Stabilizing climate requires near-zero emissions," *Geophysical Research Letters*, 35/2008, L04705, doi:10.1029/2007GL032388.

N. Nakicenovic, J. Alcamo, G. Davis, B. de Vries, J. Fenhann, S. Gaffin, K. Gregory, A. Grübler, T. Y. Jung, T. Kram, E. L. La Rovere, L. Michaelis, S. Mori, T. Morita, W. Pepper, H. Pitcher, L. Price, K. Riahi, A. Roehrl, H.-H. Rogner, A. Sankovski, M. Schlesinger, P. Shukla, S. Smith, R. Swart, S. van Rooijen, N. Victor, Z. Dadi, *IPCC Special Report on Emissions Scenarios*, Cambridge University Press, Cambridge, 2000.

P. Sheehan, "The new global growth path: Implications for climate change analysis and policy," *Climatic Change*, 91/2008, 211–231.

2. Efficient Communication†

Abstract

Society is increasingly having problems with science, and science in turn is having difficulties with society. Using the example of global climate change, climate researcher Hans von Storch and sociologist Nico Stehr show where these communication deficits stem from and make suggestions on how to get closer.

The connection between science and society does not work according to the principle of communicating pipes. The mutual observations, but also the statements and reactions of science and society pass through many intermediate stations, which, however, must not be imagined as passive switching points. The statements of natural scientists and the feedbacks of colleagues and critics are subject to multiple translations. Journeys do not always have to educate, but the passage changes the message. Neither science nor society remains unmoved. An example of this situation from the point of view of a natural scientist: When climate researchers talk about greenhouse gases having an effect on the climate, people in everyday life tend to think of poisoning the air or consuming oxygen; instead of using less energy, filters are to be put on factory chimneys. On the other hand, the scientist does not work detached from the judgments and ideas of his society and its culture. Thus, the thesis of adapting to climate change can be tantamount to an almost sinful lapse. As if one would accept the destruction of creation. As a natural scientist, one is concerned with the state, changes and dynamics of the climate, marine ecosystems or other subjects of research. But in a society where knowledge is not only the key to understanding the mysteries of nature and society but becomes the making of the world, this is a very truncated understanding of one's role.

The scientists who interact with society get into real minefields with their judgments.

They make their statements without any pressure to act. The working world of climate scientists knows about radiation, currents, storm surges and droughts, changing land use and emissions of greenhouse gases and aerosols. It does not, however, know of poverty and hunger by the millions, conflicting cultures, social conflicts, competing interests, mass unemployment, or economic recessions. The German climate scientist and the newspaper reader interested in climate are both concerned about the consequences of rising water levels in Bangladesh. Local

†This section originally appeared in von Storch, H. and N. Stehr: "Effiziente Kommunikation" [Efficient communication], *Universitas* 684: 608–614, 2003. Translated from German.

surveys on the relevant coastal problems, however, name population growth and the destruction of mangrove forests first. Only in 13th place comes sea level rise. Scientific findings focus on a world reduced to a few physical or ecological dimensions because that is the purpose of any theory. However, they encounter and work in a reality in which there is a cut-throat competition of attention and problems, in which social differences prevail and ideological attitudes have an effect, in which different forms of knowledge and perceptions are the concern of people Not only is, to cite Rudolf Virchow, "speculative" knowledge and "actually attained and perfectly established knowledge," but also pre-scientific knowledge, lay knowledge, practical knowledge, ideological knowledge, and so on part of the context. This is inevitable.

Efforts to improve the public's understanding of science through more efficient press relations and striking popularization fall far short of the mark. It is just as much about a better understanding of the public in science and an enlightened approach to the public's understanding of scientific work. An understanding of the dynamics of the socially controlled implementation of scientific expertise and the social control of knowledge, as demanded, for example, by the EU Commission in its principles of science funding, must become a natural part of science as science. Is it possible to deal with this problem constellation in a "better" way? Basic scientists should also apply their knowledge. At least they want to enlighten society, even educate it to understand scientific phenomena and developments "correctly." This becomes especially clear in the case of climate since the problem of "global warming" cannot be concretely experienced in everyday life but is a more or less plausible scientific construction. Many scientists also expect colleagues to shed their cultural baggage before sitting down in front of their computers and then stepping in front of the camera. The first demand is often and passionately made in scientific circles, the latter less often. It sounds good, but it's naive. Apart from the fact that the distinction between fundamentals and applications is an overly simplistic one, that the baggage has grown and is thus not removable, that the alleged misunderstanding is but a conflict of different forms and worlds of knowledge, natural scientists — left to their own devices — cannot adequately address the problem. They have to call in their unloved cousins: the "soft" social and cultural scientists.

What should and can they accomplish together? "Efficiency" and "resistance" are the keywords. They should help to increase the efficiency of the natural sciences and the resistance of society. Efficiency means to enable the members of society to better "understand" new, scientifically constructed ideas — which can only succeed if the "new" knowledge is placed in consistent relation to the "old" knowledge. Efficiency also means providing knowledge that is relevant to action because it is problem oriented.

Resistance has above all to do with the fact that society, its members and most important institutions accept and take into account the uncertainty and cultural conditionality of the apparently objective natural science. Resistance to knowledge worlds — natural science and the public — also means dealing with indeterminacy and indifference, exaggerations and understatements, risk perceptions and dramatizations.

The problem is, of course, that the related sciences meet with distant suspicion, know little about each other, and show an almost neurotic aversion to contact. In the eyes of the natural scientists, the social and cultural scientists are a bunch of swashbucklers who hardly know any sustainable laws, let alone seek robust insights, but offer *ad hoc* attempts at explanation that are more or less suitable for every opportunity that presents itself. In contrast, the theoretical physicist who not only determines natural temperatures with a few simple differential equations but also calculates social preferences at the same time, is impressive. Many natural scientists are convinced by rather simple, mainly traditional ideas, according to which the world is determined by climate or other geographical factors. In the past, it was race, today genetic structure. In contrast, social and cultural sciences have thrown the various "isms," especially climate determinism and biologism, onto the dung heap of ideological history with no possibility of return and explain social dynamics on the basis of social processes.

For an observer interested in state, change and dynamics, this is quite sufficient, as it satisfies the curiosity typical of the natural scientist. For someone interested in the application of scientific knowledge, distinct progress can only consist in bringing together the milieus of the estranged cousins. For natural scientists, it is not an actual problem to ask social and cultural science questions. The only problem is that then differential equations are immediately fetched from the fund of traditional methods, i.e., various kinds of determinisms dominate the further discussion. Social and cultural scientists, however, make it even easier for themselves: They declare the cousin to be incomprehensible but contextually incompetent and simply ignore her. If the natural scientist believes that the climate largely determines the life and well-being of humans and the environment, the social and cultural scientist reduces global warming to a construct of science or, at best, to a climate catastrophe whose plausibility, relevance and details no longer require further critical inquiry.

After two decades in which those responsible for German research have generously supported scientific climate research, after scientific activists have drawn public attention to the climate issue at every opportunity through the media, and after the IPCC has found its assessment, edited for public consumption, on the front pages of major newspapers, it is time to reflect on whether the current form of communication between science and the public is sufficient.

The research administration should provide incentives for social and cultural scientists to study not only the lifeworld of natural science communities inhabited by such diverse species as computer experts, geologists, meteorologists, glaciologists, and physicists. They should live in the natural science communities not only as observers and observe there the peculiarities, language, and myths, i.e., the culture of physicists, mathematicians, and geoscientists in a detached scientific way. The social and cultural scientists must engage with the residents of these alien communities. To do this, the hosts must also be willing to engage with the guests. The most important task seems to us to be to catalog the different forms of knowledge and the social division of knowledge about climate. The "modern" natural science knowledge processed by the IPCC is one form. Other ideas circulate in earth science circles. Lay people with good physical and chemical knowledge formulate their reservations in the media. In addition, there are still traditional forms of knowledge, according to which the weather has recently become less regular and predictable — apparently an anthropological constant, according to which scientific progress is *eo ipso* either positive or negative and according to which climate is nature's regulatory mechanism to put us in our place if we do too much to it. All these forms of knowledge compete for the public's attention, influence climate policy and climate perception. To be able to deal with them, we have to investigate and describe them, i.e., bring them down to a general, understandable and acceptable denominator. Another object of research is the metamorphoses of knowledge in its cycle between natural science, the public, the media, and politics. A lot happens on this transport. This is only partly a question of ignorance, of disinterest in the face of burning problems, of better-educated journalists, of lacking communication competence of scientists, of corrupt politicians and of an insufficient general education of the population. However, the study of the cycles and metamorphoses of knowledge will help to deal with knowledge — in all its forms — in a more certain, more open and more conscious way.

A particular aspect of knowledge circuits is the context-bound nature of the production of knowledge based on the division of labor. The anchoring of knowledge in social contexts is sometimes disparagingly understood as the subjectivity of knowledge. This subjectivity consists in the already mentioned knapsack of values and non-scientific forms of knowledge, but also in the — given the dominant division of labor — systematic inability to assess the situation as a whole. A physicist may understand a lot about radiative transfer, about the possibilities of remote sensing, about current methods of providing energy. His knowledge of conflict dynamics, modernization of the energy industry, the development possibilities of black Africa, the fears of the population and so on, however, is not of a scientific nature in his house, but lay knowledge. The only problem is that when he or she advises the public, politicians, or the business community in the media about the need to prevent or adapt to adverse climate changes, it is often impossible to see that two kinds of knowledge are being offered: a mélange of Nobel and common sense.

3. Climate Research and Climate Policy — Roles and Sustainability‡

Abstract

The current discussion about climate research and climate policy measures reveals a deep loss of confidence in the natural sciences. A major cause is that conditions have been created that demand a type of scientist that can be called a "scientific advocate." This type of scientist orients his research toward political and social objectives and knows how to stage himself in the media. This misguided development should be countered by reflecting on the actual tasks of the scientist. As an "honest broker," he should make his expertise available without acting with a political sense of mission. Since natural scientists are always influenced by social and cultural attitudes, it is important that they work closely with humanities and cultural scientists to fulfill their actual task. Only in this way can the institution of science, on which our society depends, survive in the long term.

In general, the scientific community agrees that human-induced processes influence the climate, that humans are changing the global climate. Climate: This is the statistic of the weather. There is widespread agreement, for example, on the following statements: The frequency distributions of temperature are currently tending toward higher values in almost all places in the world, and this trend will continue in the foreseeable future; sea levels are rising; rainfall is changing. The intensity of some extremes, such as heavy precipitation in the westerly wind belt of the mid-latitudes, will fluctuate. These changes are driven primarily by the release of greenhouse gases, i.e., carbon dioxide and methane in particular.

3.1 The scientific and the media-cultural construct of climate change

From these findings, based on actual data and observations, the scientific construct of our knowledge of anthropogenic climate change is constituted. It is widely supported in the relevant scientific circles and is particularly well received by the public, the media and politicians through the collective efforts of the *United Nations Climate Council*, the *Intergovernmental Panel on Climate Change*

‡This section originally appeared in von Storch, H. and N. Stehr: "Klimaforschung und Klimapolitik — Rollenverteilung und Nachhaltigkeit" [Climate research and climate policy — roles and sustainability], *Naturwissenschaftliche Rundschau* 63: 301–307, 2010. Translated from the German.

(IPCC). What other knowledge do we have of climate and climate change, especially that which is widespread in everyday life?

- That the climate will change by human behavior, for example by deforestation.
- That the weather is less reliable than in previous decades.
- That seasons are irregular, and storms are becoming more severe.
- That weather extremes are taking on catastrophic, unprecedented forms.

Human greed and stupidity are often blamed as the causes. For many, these are "acts of revenge by nature," that "nature is striking back." This judgmental view expresses the so-called climatic determinism (e.g., in the works of the philosopher Montesquieu and later in the studies of the influential American geographer Ellsworth Huntington), which has been of great influence in the cultures of Western societies for decades and centuries [1]. According to Huntington, the fate of man and society is largely determined by climate and weather, based on the belief that humans must live in a certain balance with the climate that suits them.

If this ideal climate changes, then the civilization shaped by certain climatic conditions is extremely endangered; entire cultures can disappear, such as Indian cultures in North America, the Mayan civilization in Central America or the Viking settlements in Greenland. This idea about climate and weather is currently primarily a *media-cultural construct*, a complex of ideas that is particularly widespread in everyday life in German-speaking countries, but also in a similar way in other parts of the West.

The newer scientific construct and the traditional media-cultural construct are competitors in the interpretation of a complex environment; two "players" in the market of knowledge. When the two forms of understanding of climate are joined, a "*modernized construct*" emerges with even greater societal impact; its scientific basis, however, becomes narrower. Public acceptance increases, the robustness of the construct against scientifically verifiable facts diminishes.

Of course, the practice of natural science (and thus its theorizing) is influenced by the media-cultural construct anyway, since scientists are bound to their own culture. Their culture conditions scientists in their understanding of their role, their interpretations guide them in their questions and in their willingness to see certain answers as argumentatively sufficient.

3.2 Postnormal science

Silvio Funtovitz and Jerome Ravetz introduced the term "postnormal science" into the discussion in the 1980s [2]. In a situation in which science must remain uncertain in its concrete, action-guiding statements, but in which, on the other hand, the statements of science are of considerable practical importance for the formulation of political decisions and in which social values and goals are affected, this form of science is being pursued less and less out of pure "curiosity." This is not immediately clear, however, because in idealistic transfiguration, curiosity is still spoken of as the innermost driving force of science. However, it is a matter of directing scientific research in such a way that its possible results meet political objectives and decisions and offer the opportunity to participate in shaping them. The focus is no longer on scientificity, methodological quality, the Popperian dictum of falsification or the Fleckian repair store of exaggerated explanatory systems [3], but on practicality. "Nothing is as practical as a good theory," it is said, referring to the expectation and willingness to help determine political decisions and guide actions. The focus is not on scientific adequacy or objective falsifiability, but on the social acceptance of scientific statements.

Climate research is currently in a postnormal situation. The inherent uncertainties of its statements are enormous, since projections into the future are required, futures that can only be represented in models and in which conditions will therefore prevail that have not yet been empirically observed. For example, we do not know exactly how the population will change or, if temperatures and water vapor content change, what will prevail in terms of Antarctica's mass balance: Increased precipitation aloft or melting at the edge. This lack of reliable knowledge has nothing to do with the incompetence of scientists, but with the poor factual situation, with the incomplete instrumental data, which span a much too short period of time for a consideration of changes on time scales of decades, with the quite problematic proxy data, which represent not only climate fluctuations, but also other conditions. Undoubtedly, there are arguments pointing to one or the other presumed answer, and plausibility considerations lead us to exclude certain developments as improbable or even impossible. But there remains a considerable residual uncertainty that will only diminish significantly over the course of years and decades.

In this situation, the representatives of societal interests pick out those knowledge claims which best support the socio-political positions favored by them and interpret them. One thinks of the Stern-Report [5] or the regular broadcasts of the US Senator James Inhofe, widely known as a "climate skeptic." But it is not only the seemingly suitable knowledge claims that are selected and put into a suitable overall frame of reference; also own new knowledge claims are formulated, so that in the end a lively collection of sometimes seemingly arbitrary claims emerges, for example, of the kind that there will be more patients with kidney stones because of global warming [6]. The scientifically untenable film *The Day After Tomorrow* is praised by public scientists as raising awareness; political and scientific achievements are lumped together by the joint awarding of the Nobel Peace Prize to Al Gore and the IPCC; and politicians disguised as professors explain to the public necessary measures as a reaction to climate change. In addition to these alarmist tendencies, there is also the skeptical counterpart found in products such as *State of Fear* by the otherwise magnificent Michael Chrichton or the film *The Great Swindle*. All this is typical for societal environments of post-normal science.

This situation cannot satisfy the historically evolved demands of natural science. It remains a discomfort that such a practice cannot be what we inaccurately call "good science," in which the argument, the critical inquiry, the clever test, the unconventional idea beyond the current paradigm reinforces progress and not the usefulness of a statement for the purpose of enforcing a policy perceived or described as correct. Even in *Science* and *Nature*, there is much that is provisional and opinionated, which stimulates the imagination and sometimes the fears of the public — and after a few years often proves to need revision. But this revision is ultimately the mechanism that pulls science out of the vortex of postnormality.

When media and public attention turns to other topics, then normal science takes hold again, and the compromises to required utility, *Zeitgeist*, and political correctness can be revised much more easily. On a small scale, we can already see this in climate research, for example, in the case of the "hockey stick," which hastily described the historical temperature development in the last millennium by an initially steady decrease, which was replaced by a dramatic, equally steady increase, or the perception of an increased storm risk, which is guided by the interests of parts of the insurance industry.

3.3 The honest broker

In view of the circumstances we have outlined, the question for the natural scientists involved is how we should deal with this post-normal situation here and now because both demands — good natural science and good advice to the public — are accepted as legitimate in science. The solution can only be that science tries to do what it can do best in principle, namely, to analyze the problem situation in a scientifically sober way. But a scientist on his own can do this only to a limited extent. The process of science is a social process; natural scientists, at least in asking questions and accepting and interpreting explanations, are not always "objective," but are also co-conditioned by their cultural understanding.

To give depth and substance to a necessary and comprehensive analysis of the climate problem, the competencies of the social and cultural sciences are needed in addition to the natural sciences. But so far, these sciences have been largely sidelined, for example, in promoting research into the social context of the climate issue, but also in the research interest shown in these social science fields in this research topic.

Occasional references to everything being socially constructed and conditioned demonstrate a clear refusal to go into the concrete, which would be necessary for a real interdisciplinary synergy of scientific efforts. But even if the overwhelming majority of social and cultural scientists still refuse a transdisciplinary approach — in the sense of a permanent, not only project-related cooperation between natural sciences on the one hand and social and cultural sciences on the other hand — to the topic of anthropogenic climate change, there are exceptional examples where the necessary accompanying social science research succeeds, such as the typology of scientists from the "Honest Broker" analysis by Roger Pielke Jr. [7].

Pielke distinguishes between five types of scientists who communicate with the public directly or indirectly in different ways and to different degrees. The "pure scientist" is essentially driven by curiosity and has little interest in seeing new findings placed in a social context. The "Scientific Conciliator" facilitates the correct understanding of undisputed scientific facts. Both types fit well with a "normal" science, which can answer questions with a high degree of certainty, and in possible social implementations, these answers are usually not controversial.

As already observed, the current climate research is not "normal," but "post-normal." Therefore, one often sees the "scientific advocate" using his or her scientific expertise not to advance knowledge in an unbiased manner, but to promote a value-based, i.e., political, agenda. The consequences of scientific insight are narrowed down to few value-consistent "solutions," or even just one. Especially the recent period has produced many scientists of this type, who work on behalf of and comment on economic or (socio-)political interests.

The fourth type of scientist, which Pielke himself clearly sees as a desirable role, has given his book its title: "The honest broker." Unlike the "scientific advocate," he expands rather than narrows the range of possible conclusions from his findings. In this way, he enables those responsible for the political process to choose, based on their interests, the "solution" that is societally desired (and not the one favored and demanded by the scientific advocate). The fifth type is the "undercover advocate," who is a "scientific advocate" according to his work, but pretends to be a mediator or an honest broker. In fact, he does neither science nor society any favors with his fraudulent labeling.

Pielke recommends that the scientific community follow the path of the "honest broker" who explains the complexity of the problems and helps to weigh the implications of possible decisions. In this way, he enables society to choose solutions to its controversies in a value-consistent and rational way, even on the basis of uncertain knowledge of contexts and possibilities to cope, for example, to deal with the prospect of anthropogenic climate change.

Another difficult situation addressed by Roger Pielke arises when policy makers fail to reach decisions because the associated policy means and goals are rejected by significantly influential sectors of society or social movements. Situations of this kind lead to the establishment of constraints in the public communication of complex problems to reduce this very complexity, in which the political system seems to feel compelled by scientific guidelines to make only one decision. In this way, politics becomes the stooge of science, and political decisions degenerate into a kind of technocratic necessity.

This is especially the case with climate policy, when the 2-degree target formulated by scientists to avoid climate catastrophe is presented as *ultima ratio*, to which politics simply must bow and which it must define as its goal. According to the rule that nothing is as practical as a good theory — since it guides action — these limited circumstances of action are politically expedient because they leave the responsibility with science and are action-relieving from the point of view of political institutions. Further politically contentious discussions are not necessary; the goals of climate policy, for example, are achieved by a matching energy policy.

It is problematic, however, that technocratically determined political decisions disappear from the publicly visible stage and from the public debate and shift into the background of the less visible scientific discussion. There, however, is just as little consensus on the practical consequences of certain objectives as in politics, and the resulting argumentative struggle among scientists becomes a politically determined dispute, which is then also conducted in the scientific community according to the rules of politics and ultimately "won" by one of the competing parties.

Politics, in turn, benefits from this process, since it is a way of making decisions that are sanctioned by the prestige of (politicized) science. This is not a sustainable use of the resource "science," whose social service of producing new knowledge and interpreting complex issues can hardly be distinguished in the public perception from the politically determined information provided by interest groups.

Pielke derives two normative requirements: Science should act as an "honest broker," i.e., it should show the full range of options for action, consequences, and risks, including unintended consequences of deliberate action. In normatively difficult situations, politics should find and implement a value-consistent "solution" — and ask science about the conditions for this.

3.4 Sustainability

Science is a societal activity, with the goal of *creating new knowledge*. Like any other social activity, it can be done sustainably or not.

What does society expect from science? That it creates knowledge to interpret a complex environment. This ability to understand the processes in the environment, one's own influence on it, the interdependencies of action and reaction, is an important contribution to the quality of life, since it means that we can actively and self-responsibly shape our lives and, in a massive way, our environment, that we can assess risks and decide more confidently.

Why is "science" trusted with this role? First of all, because of the methodology with which science works, but also because of the motives that help to determine scientific work. The methodology ensures that, as a rule, "coherent" interpretations are offered. "Coherent" here means that actions can be inferred that bring the desired result. "Incorrect" interpretations also occur, but are rare and are discovered after some time and replaced by "coherent" interpretations.

This methodology and the accompanying norms of scientific work were characterized by the science sociologist Robert K. Merton by a set of principles, which he describes as "ethos of science" [8]. Merton emphasizes, "the ethos of science is that affectively toned complex of values and norms which is held to be binding on the man of science. The norms are expressed in the form of proscriptions, preferences, and permissions" [8, p. 88]. The norms determine what should be preferred and what is still permissible.

Among the motives he characterizes as determining action and institutionalized in science is, for example, the norm of *disinterestedness*: Selfish interests should have no influence on the results of research and the behavior of the scientist. Furthermore, there is the collective norm of *organized skepticism*: Results are not subject to critical examination and falsification attempts by professional colleagues.

These normative principles do not represent an empirical descriptor of scientific behavior but are to be understood as an idealized demand that is unlikely to be realizable in pure form [9]. Nevertheless, Merton's ethos of science describes what large parts of the "customers" of science, in particular the public, expect or presuppose as a condition for the acceptance of knowledge claims. Moreover, the norms in question describe the cultural construct of the quality of scientific knowledge. These expectations regarding the emergence of scientific statements are crucial for their acceptance by the public. If these or similar normative principles are respected, it can be concluded, scientific practice can be sustained. Or more concretely: Then the public, the media and decision makers will still be listening to our current doctoral students in 20 years' time with the same attention as they are currently paying to us.

If, on the other hand, the impression arises that the principles are being disregarded, then the public's trust in the practice of science will erode, and hardly anyone will be listening to our doctoral students in 20 years' time. So how does current climate research stand in light of Merton's norms?

Disinterestedness: Selfish interests should have no influence on the interpretation of the results of research. Here, there are considerable distortions in the scientific practice of climate research.

Two camps, the "doubters" and the "alarmists," argue fiercely with and against each other, whereby the political usefulness of the statements — for or against comprehensive climate and environmental protection, for or against the precautionary principle — is in the foreground, and results that contradict these basic convictions are only conditionally recognized as "correct" by the two groups. The so far discovered "erroneous" observations in the latest IPPC report, which is supposed to document the state of knowledge, were inaccurate but publicly dramatizing statements. This can be interpreted to mean that in climate research certain political opinions have a stronger influence on the formulation of questions, the interpretation, and the selection of findings than other politically determined objectives.

Organized skepticism: Findings are subject to critical analysis and falsification by colleagues. Here, too, there are clear deficits. Gradual skepticism is accepted, but extensive skepticism is taboo and punished by exclusion from the scientific community. In the publicly discussed cases of the past months, the data material required for verification was not made available to critics; among other things, this withholding of data was justified by the fact that critics were only interested in "finding errors."

We have recently seen a clear erosion of public confidence in climate research. The weekly news magazine DER SPIEGEL, for example, asked citizens whether they were personally afraid of climate change. In 2006, 62% still answered in the affirmative; in 2010, only 42% did [10]. In the USA, Gallup asked whether the dangers of climate change were exaggerated [11]. In 2006, 30% answered in the affirmative; this year, that number rose to 48%. Of the readers of the Swedish *Aftonbladet*, after the CRU e-mails came to light, 25% rated the climate threat as a bluff, 35% as exaggerated, 17% as still unclear and 23% as very serious [12].

The erosion of public confidence evident in these figures points to changes in the perception of the importance of climate science statements and in their reception by the media, especially in light of the fact that the scientific basis of the core statements of climate science on anthropogenic climate change continue to be plausible. As already emphasized, this scientific construct of climate states: The use of fossil fuels leads to increased concentrations of atmospheric greenhouse gases; this leads to a warming of the air in the troposphere and the oceans as well as to other changes, for example in precipitation and sea level. The known trend indicates that it will continue in the future, although the magnitude of the changes can be reduced by controlling emissions of greenhouse gases.

The problem is that other statements are added to these scientifically well-founded core statements, which are formulated as "findings" — for example, about the extinction of species, the increase in hurricanes, the number of "heat deaths," population development, the relevance of democracy — but belong at best to the realm of speculation. These are interesting scientific hypotheses, which, however, are repeatedly used as politically relevant facts for argumentative purposes. The exaggerations in the report of the second working group of the IPCC (keywords: Himalayan glaciers, hurricanes and their economic damage) can be mentioned here as examples. These exaggerations, although minor in scope, contradict the principle of sustainability of scientific practice. As a result, the IPCC's presentation was perceived by many as a "bubble" that is now considered to have burst.

What can be done? The procedures described by us as sustainability must be restored; the most important element is the renewal of the institutions governed by different modes of operation. The institution of politics has to find decisions whose consequences are understandable and normatively acceptable. Science, on the other hand, has the function of clarifying how things are related and develop, independently of prevailing social values and preferences. Therefore, politics must not hide behind allegedly scientifically stringent necessities — such necessities do not exist in climate policy, nor does the goal of limiting warming to 2 degrees relative to the preindustrial state. The choice of climate targets is a legitimate political issue. On the other hand, science must not be guided in its statements by their political usefulness. The motto must therefore be: *Depoliticize science and descientize politics*.

Politics and science can in many respects be a well-cooperating pair of institutions, but a pair of collective actors with different functions, on which different expectations should be placed. We need a societal discussion about what kind of science society wants, what constitutes "good" science, and what service society expects from its "science" institution.

3.5 Options for climate policy

Climate is an important socio-political issue; it requires intensive attention by science. It must identify more options for action in climate policy. To this end, we must make climate research more open in the future — more open to questions and criticism.

In principle, there are two approaches to dealing with the climate problem: The first is to reduce the accumulation of greenhouse gases in the atmosphere, and the second is to reduce the impact on societies and ecosystems of changes in the climate due to increased greenhouse gases. The first group includes not only the reduction of emissions — commonly referred to as *mitigation* — but also global geotechnical measures to remove greenhouse gases that have already been emitted from the atmosphere. In addition to these global measures, there are also regionally limited geotechnical measures, such as changes in the urban climate through appropriate construction measures.

For a long time, climate science and climate change management practically only considered issues relevant to mitigation. The use of geotechnics was and continues to be largely taboo. In the daily lives of individuals, communities, and businesses, however, there is in any case the task of dealing with climate change impacts in a way that avoids dangerous consequences, which first requires *adaptation* knowledge.

We therefore need enabling, i.e., action-relevant, knowledge, not only about the task of reducing the release of greenhouse gases worldwide but also about adapting to the regionally different impacts of climate change.

This is necessary because significant changes in the climate are already emerging and will become even more apparent in the future. Politically demanded measures have so far had little or no success as far as emissions are concerned. The most likely scenario, the so-called "*business as usual*" scenario will prevail for quite some time, or global emissions will even increase. Even if global climate mitigation measures are successful in the coming decades, climate change will continue, affecting the well-being and development of societies. Therefore, the issue of reducing the vulnerability of people's living conditions, i.e., adaptation and precaution, must be on the agenda of public attention and political action.

The success or failure of any climate policy is embedded in social systems and their constant change. Therefore, when thinking about the future, it is not enough to estimate how the climate might change. Knowledge about the future requires knowledge about how society, its preferences and possibilities will change. Without social and cultural scientists, little success can be expected. Talking about the future must not only be done from a physical point of view.

3.6 Summary

Climate change is a serious issue; it requires our full attention. It requires scientific analysis on the one hand and political evaluation on the other. A worldview-driven perception of the dynamics, the climate consequences, the possibilities, and necessities, as served by the media-cultural construction, may generate commitment in the short term, but will hardly achieve sustainable success. Rather, a "cold" scientific analysis is needed that describes options and their consequences and, in this way, enables a normatively determined political choice.

To this end, climate science must submit to critical self-reflection as to purpose, procedure, and ethics. The subordination to a political goal and the attitude of wanting to produce knowledge suitable for a predetermined political goal must be abandoned.

It is necessary to recall the reflections of Roger Pielke and Robert Merton. Pielke recommends that science should certainly work in a problem-oriented manner and point out options for the solution of social problems, but that it should not act as an auxiliary force of social preference and try to force certain decisions. Rather, it should present all options and their consequences. Merton's principles describe the fundamental importance of contradiction, verification, openness, sustainability, independence of persons, and falsification, without which the potential of science as an action-guiding interpreter is not possible.

This will at best be achieved in outline; but even an approximation would be a success and would loosen the post-normal corset of climate science. For this, natural science is dependent on the help of social and cultural science, but also on a mature public that watches the social power factor "professor" just as closely as the cardinals, general directors, or trade union functionaries.

Literature

[1] N. Stehr, H. v. Storch: Climate works. An anatomy of a disbanded line of research. In: H. Kaupen-Haas, C. Saller (eds): *Wissenschaftlicher Rassismus — Analysen einer Kontinuität in den Human- und Naturwissenschaften*. Campus-Verlag. Frankfurt a. M., New York, 1999. – in this antology reprinted as Section 3.1.

[2] S. Funtowicz, J. R. Ravetz: Three types of risk assessment: A methodological analysis. In: C. Whipple, V. T. Covello (eds): *Risk Analysis in the Private Sector*. Plenum, New York, 1985.

[3] L. Fleck: *Entstehung und Entwicklung einer wissenschaftlichen Tatsache: Einführung in die Lehre vom Denkstil und Denkkollektiv*. Suhrkamp Verlag, Frankfurt a. M., 1980.

[4] N. Stehr, R. Grundmann: *Expertenwissen*. Velbrück Wissenschaft, Weilserswist, 2010.
[5] R. A. Pielke: Mistreatment of the economic impacts of extreme events in the Stern Review Report on the Economics of Climate Change. *Global Environmental Change* 17, 302 (2007). G. W. Yohe, S. J. Richard, S. J. Toi: The Stern Review and the Economics of Climate Change: An editorial dssay. *Climatic Change* 89, 231 (2008).
[6] T. H. Brikowski, Y. Lotan, M. S. Pearle: Climate-related increase in the prevalence of urolithiasis in the United States. *Proceedings of the National Academy of Sciences of the United States of America* 105, 9841 (2008).
[7] R. A. Pielke Jr.: *The Honest Broker*. Cambridge University Press, Cambridge, 2007.
[8] R. K. Merton: Die normative Struktur der Wissenschaft (1942). In: N. Stehr (ed.): R. K. Merton: *Entwicklung und Wandel von Forschungsinteressen. Aufsätze zur Wissenschaftssoziologie*. Suhrkamp, Frankfurt am Main, 1985.
[9] A critical analysis of Merton's norms of scientific conduct was prepared by N. Stehr: The norms of science revisited: Social and cognitive norms. *Sociological Inquiry* 48, 172 (1978); a recent discussion of Merton's criteria in the context of the Climategate affair offers R. Grundmann: "Climategate" and The Scientific Ethos. *Science, Technology, & Human Values* 38, 67–93 (2012). https://doi.org/10.1177/0162243911432318.
[10] SPIEGEL-Umfrage: Deutsche verlieren Angst vor Klimawandel, 27.3.2010. Available at http://www.spiegel.de/wissenschaft/natur/0,1518,685946,00.html (1, Mai 2010).
[11] F. Newport: Americans' Global Warming Concerns Continue to Drop. http://www.gallup.com/poll/126560/americans-global-warming-concerns-continue-drop.aspx (March 11, 2010).
[12] http://wwwc.aftonbladet.se/vss/special/storfragan/visa/0,1937,44615,00.html. The article, which was read on 1 May 2020, seems no longer available. The Climatic Research Unit (CRU) is part of the School of Environmental Sciences at the University of East Anglia (Norwich, UK). In November 2009, thousands of E-mails were made public by hacker attacks. Climate skeptics thought they had found evidence of scientific misconduct.

4. The Atmosphere of Democracy: Knowledge and Political Action[§]

Abstract

In contemporary climate change discussions among leading climate scientists, the relationship between knowledge and governance is increasingly viewed as an "inconvenient democracy." On the one hand, the discrepancy between our knowledge on climate change and citizen commitments to behavioral changes amounts to the diagnosis of an "inconvenient mind," and, on the other hand, the inertia of policies to capture progresses in knowledge leads to the diagnosis of "inconvenient institutions." The sense of political ineffectiveness felt especially among climate scientists provokes a strong disenchantment with democratic governance. As a result, it is proposed that political action based on principles of democratic governance be abandoned and replaced by top-down, centralized socio-political control. My contribution makes the case that such a view is mistaken.

Keywords: Climate change, Knowledge, Governance, Democracy, Expertise, Future present

4.1 Introduction[3]

We are well informed that the robustness and the consensus in the science community about human-caused climate change has in recent years not only increased in strength but that a number of recent studies point to far more dramatic and *long-lasting* consequences of global warming than previously thought. Although commonly referred to as "global warming," the expected consequences are increasing average global temperatures, rising sea levels and more frequent occurrences of extreme weather. Given the accumulation of greenhouse gases in the atmosphere, their retention time of hundreds or more years and, despite many efforts to reduce emissions, enhance resilience and implement new technologies, the relationship

---

[§]This section originally appeared in Stehr, N.: "The atmosphere of democracy: Knowledge and political action," pp. 69–91 in J. Glückler, G. Herrigel, and M. Handke (eds.), *Knowledge for Governance 2020. Knowledge and Space.* Vol. 15. Cham: Springer, 2020.

[3]My discussion of the relation between knowledge, expertise, and democracy draws on a couple of earlier reflections, for example, Stehr (2015, 2016b). I am grateful to Michael Handke for his comprehensive and constructive review of my manuscript. I thank Scott McNall for his helpful comments.

between climate and society is bound to change in novel and unpredictable ways (see Stehr and Machin, 2019).

Under the circumstances, how is it possible, many scientists now ask, that robust science-based evidence does not motivate and encourage major political action in all societies and changes in the conduct of civil society members around the world? How is it possible that democracies in particular have done so little to effectively combat the risks of climate change and simply failed to attend to the dangers of global warming?[4] After all, the nature of the future present is very much predicated on decisions taken now.

Being disenchanted with the workings of democracy and blaming democracy for a host of social, economic and political ills is not a new complaint: "Lamenting the failings of democracy is a permanent feature of democratic life, one that persists through governmental crises and successes alike" (Runciman, 2013a). However, the reference "climate change" is a novel reason for a fundamental concern about the fate and future of democracy.

4.1.1 *Blaming democracy*

Climate scientists, social scientists, and the media as well as environmental activist groups (NGOs) concerned with climate change refer to a "future present" of exceptional circumstances[5] and protest that "evolution did not design us to deal with such problems" (Jamieson, 2014:61; Di Paola and Jamieson, 2018).[6]

[4]I am using the concepts "risk" and "dangers" not as overlapping terms but in the sense in which they were introduced as opposite concepts by Niklas Luhmann (2005:23). The risks of climate change can be attributed human-made decisions while we are exposed to the dangers of climate change. An example of risk-taking decisions related to climate change can be studied in today's State of California: People are moving into high fire-risk zones, i.e., the population of California grew by 3 million between 2000 and 2010, and, "in 2017 over a quarter of the state's population lived near moderate or high-risk fire corridors. With this increase in population comes a higher possibility of a human-made wildfire. And as people move into these high-risk areas, more buildings are in harm's way: Structures generally burn longer than vegetation, allowing fire more time to spread" (cf. *The Guardian*, November 11, 2018, "Why are California wildfires so bad?", https://www.theguardian.com/world/ng-interactive/2018/sep/20/why-are-california-wildfires-so-bad-interactive.

[5]The useful concept of a "future present" is Niklas Luhmann's (1976:140) terminology: "If we characterize processes or activities as beginning or ending, we use a terminology which belongs to the present. If we use these expressions to refer to distant dates — for example: The Roman Empire began to fall — we refer to a past present or to a future present."

[6]An incessant amplification in the discourse of imminent threats, many may recall the 1986 SPIEGEL title with Cologne cathedral under water, can paradoxically turn out to be supportive of

Members of the same groups assert impatiently that no one is listening to the diagnosis of historically unprecedented risks and dangers.[7]

In important respects, therefore, the discourse of climate science having established the fact of anthropogenic climate change has become by necessity forward looking. The focus has shifted to how will it be possible to govern societies in the not too distant under the massive impact of global warming. How will it be possible to govern a future present that is anticipated to be altogether different from the societal context in which democratic systems originated and flourished in the past? In cases which I will identify, strong opinions promoting the need to suppress political liberties in the wake of profound future environmental changes are not unusual anymore yet have not received systematic attention in social science.

In this essay, therefore, I will bring this disenchantment with democracy, especially in its currently dominant liberal version under the spotlight. My essay is about the struggle to align politics and policy with science. I will critically probe the argument that policy makers are going to have to act, even without a broad public mandate and legitimacy. Time is very short before the future with disastrous damage is locked in. But rather than lamenting about the inconvenience of democratic governance, it is important to reflect upon ways of enhancing democracy, not despite but *especially* in light of the massive challenges of a changing climate. Coping with major environmental challenges is best accomplished, as history shows and as I will argue within the bounds of democratic rather than authoritarian political systems. The essay reframes our changing climate as an issue of political governance and not as merely an environmental or as an economic issue.

I will advance my argument in a number of steps: I will address, first, the growing assertion that contemporary democracies face exceptional circumstances. Second, I will reflect on the classical and present-day social science discourse on the erosion of the foundations of democracy. Third, I will describe the growing sentiment of an *inconvenient democracy* among climate scientists, other scholars, NGOs and the media. Climate scientists propose to overcome the inability of modern democracies to cope with the disastrous consequences of climate change by abolishing democracy. The alternative of course is to strengthen

---

the opposite virtue, namely, as a defense of the present and encouraging skepticism toward scenarios of impending dangers. This represents a psychological mechanism not unlike the everyday attitude toward weather extremes widely interpreted as an affirmation of the normal course of climate (cf. Stehr, 1997; Stehr and Machin, 2016b; Stehr and Machin, 2019).

[7]As Bill McKibben (2018) for example notes: "Over and over we've gotten scientific wake-up calls, and over and over we've hit the snooze button. If we keep doing that, climate change will no longer be a problem, because calling something a problem implies there's still a solution."

democracy. Forth, I will consider the proposed shift in role for climate scientists as policy makers. In the final section, I will examine the serious deficiencies in the assertion of contemporary society as an "inconvenient democracy."

4.1.2 *The rise of "exceptional circumstances"*

As never before, the continuity from past to future is broken in our time.

Niklas Luhmann ([1992] 1998:67)

In the past, war-like conditions and major disasters typically were seen to justify the abolition of democratic liberties, if only temporarily. The present appeal to exceptional circumstances from the critics of dominant government climate policies around the world, echoes this sentiment, demanding the elevation of a single socio-political purpose to ultimate political supremacy.[8]

With climate change we are confronted with a historically novel situation and future present: Climate change within historical times is locked in. Most of the scientific discourse has been devoted to establishing the phenomenon that there is anthropogenic climate change. The issue that climate change is anthropogenic has been settled, and it has become clear that unless increasingly vigorous political, economic, and societal measures are implemented, the planet will continue to experience warming "greater than it has been for more than half a million years" (Nordhaus, 2013:325). What is not settled in science is a range of important questions such as the speed of global warming or the nature of the consequences of climate change on various significant attributes of human existence and in different regions of the world.

Governing the consequences of climate change refers to a time scale and anticipated societal transformations that are clearly beyond human imagination and current political institutions. Except for reference to singular historical events, for example, war, revolution, economic collapse or the struggle for national liberation, there are no large-scale human experiences within historical times to which the claims of the climate science community can appeal as it begins to reflect about a "future present" in which massive impacts of climate change have set in. This relates to all levels within society and its relations abroad, for instance, how the world makes and uses energy, the virtue of the nation state, migration patterns, the global economy, and civil societies. In such contexts, crisis conditions promote

[8]For a discussion of *exceptionalism* in political theory, critical security and citizenship studies, see Best (2018).

the creation of emergency powers, the delegitimation of the previous political order, the abolition of liberty and justice and the installment of revolutionary governance. Although the past is by no means a foolproof guide to the future. It is, however, often the only guide we have.

Appeal is therefore made to extraordinary circumstances or a war-like footing (Lovelock, 2009; McKibben, 2016) that necessitates the suspension of freedoms and the political ascent of climate scientists. As the French political scientist Pierre Rosanvallon ([2011] 2013:184) stresses: The central nation state is seen as the only source of security in the face of radical risk. It is the hope that an appeal to extraordinary circumstances, i.e., to a threat to the very existence of civilization if not humankind "alone might be able to give capacity and [...] energy back to a failing or hampered [political] will." Frank Fischer (2017: 54) complements this in criticizing that "current political-economic efforts on part of contemporary democratic systems to deal with problems such as global warming [... are] little more than limited symbolic gestures, especially given the pressing constraints of time." The problem of global warming and its consequences does not merely pertain to contemporary democratic governance and a missing commitment of citizens to change their ambitions and behavior. Above all, a future perspective is needed (Lovelock, 2009). The future perspective imposes its own norms on the present (cf. Jonas, [1979] 1984:143).[9]

But how does one govern well under exceptional circumstances? This question encounters two countervailing forces: That of an *inconvenient mind*[10] and of *inconvenient social institutions*. The former relates to a public that is assumed to be "present-centric" (Skidelsky and Skidelsky, 2012:130), i.e., comfortable with the status quo, and that justifies imposing one's own (superior) ideas on that of future generations of citizens (because do we really need to care whether the future public cares?). The latter relates to a strong state in the form of a command society. In other terms, good governance of society based on citizen participation

[9] Hans Jonas ([1970] 1984:143) interrogates the Baconian idea (executed for example within Marxism) of dominating nature by increasing the power of human over nature in his search for an ethic of the technological age. Jonas designates the Baconian ideal as the source of an ethic that aims predominantly at the future and therefore imposes its norms on the present.

[10] The reference to the inconvenient mind is of course a play on word using the better-known metaphor "an inconvenient truth" as its anchor. A fairly straightforward example for an inconvenient mind in the case of climate change is to suggest that the science of climate change is much too complicated for the average citizen to comprehend. A less "neutral" version of the inconvenient mind would be to suggest that the public is intellectually incapable of grasping the idea of global warming and its consequences.

must be subordinated by almost any means to the defeat of the exceptional circumstances.

It is the single purpose of defeating the exceptional circumstances that legitimizes the temporal suspension of liberties (Hayek, 1944:189). However, is any massive absorption of powers in the hand of the state and its representative's reversible, in the long run? And, are the potential consequences of climate change the equivalent of (abrupt) war-like conditions? How can one pinpoint the onset of exceptional circumstances?

The deficiencies of democratic governments are many and by far exceed the issue of climate change and its societal consequences; but is it therefore justified to reach such a disparaging conclusion as is the diagnosis of an inconvenient democracy? After all, authoritarian and totalitarian governments do not have a record of environmental accomplishments; nations that have followed the path of "authoritarian modernization/environmentalism" such as China or Russia cannot claim to have a better record.[11] Nonetheless, the disenchantment with democracies continues to be advanced and perhaps is becoming even more vocal as entrenched climate policies fail to live up to their promise.

4.2 Inconvenient democracy

The assertion of exceptional circumstances and its concomitant promotion of the need to overcome an "inconvenient democracy" derives its intellectual sustenance from a range of new and classical considerations and they lead to different forms of blaming with different addressees.

4.2.1 *The erosion of democracy: The classical perspective*

In the classical social science literature, the threat to democracy that issues from an uneven access and distribution of knowledge in societies, for example, in social inequality formation in society (see Stehr and Machin, 2016a) has in the eyes of many observers radically displaced earlier, optimistic Enlightenment views regarding the resilience and even the possibility of a democracy based on a general circulation of knowledge in society.[12] Numerous authors, from Max Weber to

[11] As Bruce Gilley (2012:287) explains, "Authoritarian environmentalism" refers to "an emerging theory of public policymaking in the face of severe environmental challenges. It has been discussed both as a prescriptive model of how countries should effectively respond to such challenges, and as a descriptive model of how they are likely to respond."

[12] There is good reason to be skeptical toward the idea that either the notion or the realities of the knowledge gap or the information overload, however defined, are genuinely new. One has only

Robert Michels, have explicated these and other threats to representative democracy.

Given the unstoppable advance of bureaucracy in modern societies, Max Weber ([1918] 1994), for example, feared a kind of *pacifism of social impotence* of the citizenry, for in the face of a "growing indispensability and hence increasing power of state officialdom [...] how can there be any guarantee that forces exist which can impose limits on the enormous, crushing power of this constantly growing stratum of society and control it effectively? How is democracy even in this restricted sense to be *at all possible*?" (Weber, [1918] 1994:159).

Robert Michels ([1915] 1949) in his classical study of the undemocratic tendencies in the social democratic party, a political organization that actually aspires to and fights for democratic goals, refers to an almost "natural" state of incompetence and immaturity of the mass of people in modern democracies. And since those of rank and file are incapable "[...] of looking after their own interests, it is necessary that they should have experts to attend to their affairs" (Michels, [1915] 1949:93). Seldom is the rank and file willing to throw off the authority of the expert leaders and dismiss them from control.[13] Numerous of the classical concerns about the viability of democratic governance find the echo in contemporary reflections about the fragility of democracy.

4.2.2 *The erosion of democracy: The modern perspective*

A deep-rooted pessimism about the psychological make-up of human beings; the temporality of human thought; the failure to mobilize individuals for the cause of effective climate policies; the inability of government given constitutional constraints to attend to long-term goals; the fragility of the political order; the influence of vested interests on the political agendas of the day; and in the case of anthropogenic climate change, the addiction to fossil fuel; last but not least, the

---

to refer to the convergence of societal diagnoses proposed, at the dawn of the last century, by Georg Simmel, Sigmund Freud and Walter Benjamin, among others, of a cultural age displaying severe overstimulation, discontinuities and overload.

[13]Whether the disillusioned conclusion Robert Michels ([1915] 1949:95) draws in light of the tendencies he observes, namely that "social democracy is not democracy, but a party fighting to attain democracy" is inevitable, i.e., universally applicable as a kind of iron law, surely is of course contestable, although many observers are prepared to concede that Michels' has discovered one of the few law-like relations in social science. For more recent studies by economists, sociologists and political scientists who take Michels' challenge about the inevitability of oligarchic tendencies in organizations on board see Williamson (1975, 1985, 1995), Granovetter (1985), Foucault ([1981–1985] 2005) and Stehr and Adolf (2018:21–324).

ineffectiveness of the climate science community itself to insure that their message does not fall on deaf ears.[14]

4.2.3 *Blaming the people*

Daniel Kahneman sums up the growing skepticism regarding citizen motivation when he states:

> *the bottom line is that I'm extremely skeptical that we can cope with climate change. To mobilize people, this has to become an emotional issue. It has to have the immediacy and salience. A distant, abstract, and disputed threat just doesn't have the necessary characteristics for seriously mobilizing public opinion* (cited in Marshall, 2014:57).

The mass of citizens, it seems, simply cannot be won over to endorse and follow the course of scientifically based policy options. The large majority of citizens is basically inclined to act irrationally (cf. Schumpeter, 1942:262–263). The climate scientist Hans-Joachim Schellnhuber (2011:29)[15] gloomily relates why climate change communication does not reach civil society, "my own experience and everyday knowledge illustrate that comfort and ignorance are the biggest flaws of human character. This is a potentially deadly mix." However, to view democracy and politics in terms of the competence of the individual citizens is to argue in favor of a micro-sociology without a macro-sociology. The reference to the public perceptions of science and expert knowledge goes beyond the implicit or explicit assumption that the public has basically deficient information and knowledge, is perhaps even reactionary and tends to respond to complexity with trepidation (cf. Gauchat and Andrews, 2018).

The apparently widely shared ability to avoid knowing what the future could bring can of course also be interpreted as a psychological "incentive" to live with the knowledge about the limited knowledge on the outcome of events that are located in the future (cf. Gigerenzer and Garcia-Rettamero, 2017). Meanwhile,

---

[14] Efforts in climate change communication are predicated on the conviction that if the public only knows the facts about climate change and begins to understand just how serious the problem is, they will raise their voices and demand that our governments and corporations do something (cf. Andrew Revkin, http://dotearth.blogs.nytimes.com/2014/04/16/a-risk-analyst-explains-why-climate-change-risk-misperception-doesnt-necessarily-matter/?_php=true&_type=…).

[15] The climate scientist Hans Joachim Schellnhuber in an interview with DER SPIEGEL (issue 12, 21 March 2010, p. 29) in response to the question on why the messages of science do not reach society.

political scientists, who in many ways, for example, have been concerned about the voters' lack of information, have begun to stress that the democratic-political system works in spite of citizens' ignorance (Kuklinski, 1990). Or, as Petersen and Aarøe (2013:289) have more recently documented, despite the widespread lack of extensive political knowledge, "citizens readily form opinions on what constitutes the best and most efficient policies."

An appraisal more in support of the political virtue of knowledgeable citizens is advanced by Seymour Martin Lipset and his colleagues ([1956] 1962): Lack of information, passivity, and lack of interest of rank-and-file members in the affairs of an organization is in the interests of the powerful and supports their capacity to perpetuate power advantages. It seems that it is not so much the volume of knowledge or information that citizens command that impacts the relation between democracy and knowledge, but rather the importance of democracy-enhancing individual and collectively shared *value-orientations*; or, as Robert Dahl (1977:1) argues: It is "the ways in which we think about ourselves as a people," that support the existence and the stability of democracy. Of course, value-orientations and educational achievement are connected: "Education presumably broadens men's outlook, enables them to understand the need for norms of tolerance, restrains them from adhering to extremist and monistic doctrines, and increases their capacity to make rational electoral choices" (Lipset, 1959:79).

4.2.4 *Blaming the political class*

In the eyes of many from the climate science community, not only citizens but also politicians are not ready to pursue policies that effectively address climate change: Climate activist, climate scientists, some politicians and many other observers agree that the recent climate summits in Copenhagen, Cancun, Durban and Warsaw were failures. The summits did not result in a new global agreement to cope with the emissions of greenhouse gases. The subsequent 2015 Paris agreement, widely regarded as a historical achievement, seemingly marks a general scientific and public consensus that anthropogenic climate change is a very serious threat to human civilization and its environments. The treaty, however, is non-binding. There are no formal sanctions if a country should fail to live up to its commitments regarding the efforts in terms of mitigation, adaptation or finance, and there is no guarantee how far reaching the Paris agreement will be. This problem came to the fore on 1st June 2017, when the United States, led by

President Donald Trump, announced their formal withdrawal from the treaty rejecting the scientific consensus that greenhouse gas emissions are warming the planet.

Although under the terms of the Paris agreement, the U.S. cannot formally begin the process of withdrawal until November 2019, the current administration is already embarked upon a strong anti-environmentalist agenda. In his announcement, Trump was fulfilling his campaign pledge to "end the war on coal" and his purported aim to reclaim sovereignty for the American people and put "America first." But as has been much remarked upon, shortly after his announcement in which Trump emphasized that he was elected to represent "the people of Pittsburgh and not Paris," the mayor of the state of Pittsburgh, Bill Peduto, voiced his criticism of the withdrawal and proclaimed the state's commitment to the treaty. Indeed, a number of American states and cities will continue to follow its announced climate policies, offering "a profound counter to Trump's anti-environmental crusade" (Bomberg, 2017:5). What this scenario illustrates is the high degree of politicization of the issues of climate and climate change in the contemporary world.

The nature of the relation between temporality and democracy indeed justifies doubts about the effectiveness of democratic governance in the face of longer-term future risks and dangers of climate change. Issues of temporality refer to at least a couple of significant matters driven by distinctive but related systemic conditions of democratic governance: On the one hand, democratic governance is captivated by the *immediacy* of frequently changing "events" that come and go often rapidly, as much as it is affected, on the other hand, by constitutional rules of representation that prescribe relatively short frames of *temporality*. The public perception of the urgency of political issues is dynamic and relative. The attention that is given to climate change very much depends on the perception of the importance of other political issues at any given time, especially on the perception of pressing economic issues.

Are democracy and societal institutions constrained by short-term constitutional frames and governed by principles of liberty, such as the market, capable of dealing with harms and risks to society that are located in the future? How can democracies sustain interest in a future present that is a couple of decades away and thereby escaping the typical media issue attention cycle (Downs, 1972; McDonald, 2009) of events?

There is a parallel discourse in social science to which I now turn that expresses strong doubts about the "sustainability" of modern democracies. It highlights a crisis that is not only triggered by major environmental problems but also by various structural and secular challenges faced by present-day democratic governance.

4.2.5 *Are democracies dying?*

The discussion in the climate science community about the inadequacies of democratic governance converges with assessments of the present state and future of democracy in the social sciences. It was only a few years ago that political scientists proclaimed the end to history (Fukuyama, 1992) and with it the ultimate victory of democracy. Today, political scientists — Francis Fukuyama (2018) included — are much more likely contemplating the dissolution of democracy. *Even* titles like "The Future of Freedom" (Zakaria, [2004] 2007), "The Retreat of Western Liberalism" (Luce, 2017), "How Democracy Ends" (Runciman, 2018), "How Democracies Die" (Levitsky and Zinblatt, 2018), "The People vs. Democracy" (Mounk, 2018) and "Can Democracy Survive Global Capitalism?" (Kuttner, 2018) give an indication of it. The dispute about climate change and climate policies plays a central role in the contemporary shift of the debate about the well-being of democracy. In response to multiple societal changes underway, the argument concludes, democracy loses its legitimacy in the eyes of its citizens.

The conclusion of social science observers must therefore be that contemporary democracy — in many ways whether by design or as the outcome of structural economic, political and moral changes — is on its way to autocratic forms of governance. For instance, the erosion of democracy manifests itself in processes of depoliticization, the substitution of politics by techniques of management or the restriction of the public sphere or (cf. Rosanvallon, 2006:228; also Swyngedouw, 2011): "in a hollowing out of citizenship, the marketization of the public sector, the soul-destroying targets and audits that go with it, the denigration of professionalism and the professional ethic, and the erosion of public trust" (Marquand, 2004:172). Democratic governance is increasingly muted by the rapid abolition of democratic principles of political equality, and even replaced by autocratic forms of governance that echo Robert Michels ([1915] 1949) century-old iron law of oligarchy.

What distinguishes the discussion about the poor health of democracy among social scientists and climate scientists is the remedy that both sides advocate: On the one hand, social scientists discuss efforts that could restore democracy, for example, rebuilding "a society of similar individuals" (Rosanvallon, [2011] 2013) through the active participation of a large number of citizens that will shape the agenda of public life. On the other hand, climate scientists and other observers of global climate change disparage about the very capacity of democratic governance in coping effectively with the large-scale environmental problems and therefore call for a more authoritarian state and/or a state where decision making by technical experts is given weight. But then democracy is allegedly dismantling itself.

Colin Crouch (2004:4), for example, describes the transition of democracy to post-democracy in the following terms: "Under the conditions of a post-democracy that increasingly cedes power to business lobbies, there is little hope for an agenda of strong egalitarian policies for the redistribution of power and wealth, or for the restraint of powerful interests."

Post-democracy is also accompanied by the swift erosion and disavowal of democratic rights and values, as Richard Rorty (2004:10) argues: "[A]t the end of this process of erosion, democracy would have been replaced by something quite different. This would probably be neither military dictatorship nor Orwellian totalitarianism, but rather a relatively benevolent despotism, imposed by what would gradually become a hereditary nomenklatura." In some of the images of "post-democracy" as a state of the state — a return to aristocratic society — has already been achieved: Self-appointed elites claim to carry out the wishes of the masses.[16] In short, as Pierre Rosanvallon (2006:228) emphasizes, politics has been replaced "leaving room for one sole actor on the scene: International society, uniting under the same banner the champions of the market and the prophets of the law." This marks a political development that is very much welcome by representatives of the climate science community.

The radical conclusion drawn by some observers especially by those who favor and promote the role of experts and expertise as a form of enlightened leadership is that democracy itself is inappropriate, that the slow procedures for the implementation and management of specific, policy-relevant scientific knowledge lead to massive, unknown risks and dangers. Civilization-as-we-know-it may come to an end. Assuming it is not already too late, appropriate environmental governance has to look very different. To create a globally sustainable way of life, we immediately need, in the words of German climate scientist Hans Joachim Schellnhuber (cf. WGBU, 2012), a "great transformation." Part, if not the core of the required great transformation appeared to be a new political regime and forms of governance: For example, as expressed by the Australian scholars David

[16] Hans Jonas' ([1979] 1984:147) sober response to such a claim is quite appropriate and worth citing in this context: "if [...] only an elite can assume, ethically and intellectually, the kind of responsibility for the future which we have postulated — how is such an elite generated and recruited, and how is it invested with the power for its exercise?"

Shearman and Joseph Wayne Smith (2007:12) in their book *The Climate Change Challenge and the Failure of Democracy*: "We need an authoritarian form of government in order to implement the scientific consensus on greenhouse gas emissions." In the same vein argues Mark Beeson (2010:289) when he brings into play the notion of good authoritarianism: "[...] given the unprecedented and unforgiving nature of the challenges we collectively face [...] forms of 'good' authoritarianism, in which environmentally unsustainable forms of behavior are simply forbidden, may become not only justifiable but essential for the survival of humanity in anything approaching a civilized form." Another proposal is for a distinctively political role of climate scientists. In most countries' climate science is successful in equipping governments with the authority of the correct point of view about climate change. However, climate science fails to ensure that governments act on the authority of science.

What is the alternative? One alternative is an exchange of leadership and the rule of the knowledgeable class. The idea to exchange political leadership is not only to put science and scientists at the center of governance but also to depoliticize the issue of climate change (cf. Swyngedouw, 2010; Aitken, 2012).

4.3 Enlightened leadership?

Within the broad field of climatology and climate policy, one is able to discern growing frustration with the virtues of democracy and a mounting appeal to exceptional circumstances and the promotion of the role of scientists and experts in policy making. The impatience with democracy and the shifting understanding of the role of scientists can be observed with a change in the function of the *International Panel of Climate Change* (IPCC). The IPCC no longer considers itself a scientific organization with the mandate to offer alternative policy options for political discussion and decision but as a body of experts demanding that options for political action it identifies be rapidly realized.

Robert Stavins, the director of Harvard's Environmental Economics Program and a co-author of the IPCC Working Group 3 report notes a "bottom-up demand which normally we always want to have and rely on in a representative democracy, is in my view unlikely to work in the case of climate change policy as it has for other environmental problems [...] It's going to take enlightened leadership, leaders that take the lead."[17]

[17] As quoted in Andrew Revkin, "A risk analyst explains why climate change risk misperception doesn't necessarily matter," *New York Times*, 16 April 2014.

The social scientist Evelyn Fox Keller (2017:107) makes the strong case for an immediately effective, practical political role of climate science, given the seriousness of the problem of global warming:

> *there is no escaping our dependence on experts; we have no choice but to call on those (in this case, our climate scientists) who have the necessary expertise [...] Furthermore, for the particular task of getting beyond our current impasse, I also suggest that climate scientists may be the only ones in a position to take the lead. [... and] given the tacit contract between scientists and the state which supports them on the other, I [...] also argue that climate scientists are not only in a position to take the lead, but also that they are obliged to do so.*

4.4 Science, knowledge and democracy

The strong desire to reach specific policy outcomes spelled out by the climate science community leads many to believe that scientific knowledge is somehow immediately performative or is an immediately persuasive form of knowing. Such a conception of knowledge privileges knowledge as a policy instrument by ignoring the limits of the power of knowledge (Stehr, 1991; Prewitt, 2010; Sarewitz, 2010). On this doubtful basis alone, it is not surprising that climate scientists at least sympathize with the suspension of democratic process.

However, there are a number of plain weaknesses of the inconvenient democracy position that I will enumerate now in some detail. My observations are organized into five counterarguments:

First, and importantly, one encounters a flawed understanding of scientific knowledge and its potential role in political contexts. Scientific knowledge neither is immediately performative (knowledge equals control and represents practical reason) nor is it immediately persuasive (i.e., knowledge convinces unencumbered). Knowledge alone does not generate a profit or score goals (cf. Van Dijk, 2014). One of the fundamental flaws in the portrait of an inconvenient democracy is the failure to recognize the social character of knowledge in general and the contested and often ambivalent nature of political knowledge in particular. Recognizing the proper function of knowledge assures a premature political closure, i.e., the depoliticization of the issue of climate change and climate policies.

It is more appropriate to characterize knowledge not as *something that is so* but as a generalized *capacity to act* on the world, as a model *for* reality, or as the ability to set something in motion (Stehr, 1994; Grundmann and Stehr, 2012; Stehr and Adolf, 2018). The German term that best describes knowledge as a generalized capacity to act would be *Handlungsvermögen*. The verb *vermögen* signals "to be able to do," while the noun *Vermögen*, in this context, is best translated as "capacity" (rather than "fortune" or "wealth").[18] The capacity to act — the ability to put something into motion — extends to the capacity to generate "symbolic action." For example, symbolic action may involve the ability to formulate a hypothesis, carry out a ritual, find a new metaphor for an established term,[19] assess "facts," organize the literature on a topic or defend a thesis against "new facts." The capacity to act, in other words, refers not merely to the possibility of accomplishing something in terms of a material and physical performance such as, for example, to make fire, to drive a car, etc. Capacities to act also refer to *intellectual* abilities as well as the production of *meaning* such as may be found in the detailed description of the bundle of skills that I call *knowledgeability* (cf. Stehr, 2016a). This is most likely also the reason why Norbert Elias (1984:252) defines knowledge as "the social meaning of human-made symbols, such as words or figures, in its *capacity as means of orientation*" (my emphasis).

Knowledge, as a generalized capacity for action, acquires an "*active*" role (i.e., is put to work) in the course of social action only under certain circumstances, namely where social action does not follow purely stereotypical (effortless) patterns (Max Weber), or is strictly regulated in some other fashion. Under conditions of ritualized social conduct, a break in the continuity between past and future is not happening. Past and future are securely looked in through taken-for-granted sequences of events.

Niklas Luhmann's observations about the conditions for the possibility of making decisions in the first instance perhaps allow for an even broader understanding of the use of knowledge but also confirm my description of the likely usefulness of knowledge only under conditions of degrees of openness of the circumstances of action. Decision-making, Luhmann ([1992] 1998:67) writes "is possible only if and insofar as what will happen is uncertain."

The circumstances of action that I have in mind may also be described as the capacity of actors to alter or stabilize a specific reality. However, the capacity *to get things done*, to alter and affect reality, as well as the ability to intervene in a context that otherwise would change, is not symmetrical with the capacity to act (knowledge). Knowledge and control should not be symmetrical: "Foresight and control is highly fragile in reality, it can be shown that a persistent progress of knowledge neither leads necessarily to an improvement of foresight nor to an

[18] Georg Simmel ([1907] 1989:276), in his discussion of money as a generalized code, uses the concept Vermögen to describe the fact that money is more than merely a medium of exchange; his definition of money thus transcends a merely functional understanding of its social capacities.

[19] I refer in this context, for example, to Donald Schon's ([1963] 1967) reflections in *Displacement of Concepts* (cf. also Haldane's, [2009] 2013).

improvement of control" (Tenbruck, 1977:223). The ability to do something is dependent on the control over the conditions of action. The lack of control over the political conditions of action is an apt description of the societal role that fits the position of climate scientists today and continues to be the case as long as they have not appropriated political power.

Second, one leading assumption of the climate science critics of democracy is a misunderstanding of the climate problem and a misleading framing of the policy process.[20] The result of this misunderstanding of the climate problem and of the climate policy process is a fundamental framing error, representing *climate change as a conventional environmental "problem" that is capable of being "solved."* It is neither of these.

Rather than being a *discrete* problem to be solved, climate change is better understood as a persistent condition that must be coped with and can only be partially managed more — or less — well. The climate issue is one part of a larger complex of such conditions encompassing population, technology, wealth disparities, public values, resource use, etc. Hence it is not straightforwardly an "environmental" problem either. It is axiomatically as much an energy problem, an economic development problem or a land-use problem, and may be better approached through these *multiple avenues* than as a problem of managing the behavior of the Earth's climate by changing the way that humans use energy.

This makes climate change a "*wicked*" problem.[21] A wicked problem is the impossibility of giving the policy issue a definitive formulation: The information needed to understand the problem is dependent upon one's idea for solving it. Furthermore, wicked problems lack a stopping rule: We cannot know whether we have a sufficient understanding to stop searching for more understanding. There is no end to causal chains in interacting open systems of which the climate is the world's prime example. *Climate change policies are best embedded in comprehensive policy perspectives* that attack climate change *indirectly* accepting for example that decarbonization will only be achieved successfully as a benefit contingent upon other goals which are politically attractive and pragmatic.

Third, in a related manner, the dominant political approach concentrates almost exclusively on a *single effect* that governance ought to achieve, namely a reduction of greenhouse gas emissions and perhaps necessary measures of

[20] My critique of the dominant framing of the climate problem draws on our *Hartwell Paper* (Prins *et al.*, 2010).

[21] Wicked problems are embedded in multiple social systems. Originally described by C. West Churchman (1967) and later explicated more comprehensively by Horst Rittel and Melvin Webber (1973; cf. Peters, 2017) in the context of urban planning, wicked problems are issues that are often formulated as if they were susceptible to a simple, unilinear solution when in fact they are not.

adaptation to climate change. In doing so, it excludes other, more complex forms and conditions of action. By focusing on the goals of political action rather than its conditions, the contentious issue of climate change is reduced to scientific or technical issues. Sociopolitical issues are neglected. The politicization of climate science leads to a depoliticization of climate changes. Matters relevant to the public are permanently taken out of politics (see also Jasanoff, 2012).

Equally deficient in this context is the focus on a *single approach* to attack climate change, namely a reduction of greenhouse gases, especially CO_2. The exclusive framing of climate policy directed toward a reduction of emissions ignores what Roger Pielke Jr. (2010) calls the "*iron law*" of climate policy. The iron law merely states that while people are often willing to pay a certain price for environmental policy goals, their willingness has its limits. The exact limit varies of course from place to place and household to household. The massive resistance of the "Yellow Vest" protests in France against a fuel tax increase on a regular basis intended by the French government to fight global warming in the early winter of 2018 is a perfect example of Pielke's law. The protests of the Yellow Vest movement forced the Government to cancel the tax increase. Public support for climate policies declines as a function of the impact of such policies on the household costs. A convergence of ecological and economic policies is not impossible. However, such a convergence likely tilts toward the economic part of the equation when emission reduction policies collide with economic growth or labor market policies.

Fourth, the generally pessimistic assessment of the ability of democratic governance to respond to, cope with and control exceptional circumstances is linked, if only implicitly, to the then peculiar optimistic assessment of the potential of large-scale planning in the sense of social engineering. Planning on any scale is hardly straightforward. Not only the capacity of governments but also the general possibility to plan for the future present of societies is rather limited, perhaps absent (see Tenbruck, 1977:138). Economic and social planning conceptions widely discussed in the affirmative decades ago have fallen into disrepute (see Giddens, 2009: 94–100). Certain schemes to improve the human condition have failed; as is demonstrated case-by-case by James Scott (1998) in his book "Seeing like a State." The once active academic program of and enthusiastic support for Futurology about desirable futures has vanished (Seefried, 2015). Modern decentered, functionally differentiated societies preclude de-differentiated, society-wide social planning in principle (Luhmann 1976, [1992] 1998).

Fifth, in the reasoning of the impatient critics of democracy, one notes an inappropriate fusion of nature and the nature of society. The uncertainties (related to climate) that the sciences of the natural processes claim to have eliminated and the authoritative consensus that the sciences have thereby acquired, are simply transferred to the domain of societal processes. Consensus on the evidence, it is argued, should motivate a consensus on political action. What becomes desirable is a rational design of social order "commensurate with the scientific understanding of natural laws" (Scott, 1998:4), for example, a comprehensive engineering of human settlement and production. Designing society top down is schematic and ignores the essential realities of any really existing social order: The constitutive uncertainties, fragility and complexity of social, political and economic events, the difficulty of anticipating the future present is treated as minor obstacles that can be encircled as soon as possible — of course by a top-down approach — by implementing policies that the faith in scientific knowledge prescribes. This undermines the dignities, pluralities and conflicts that are immanent features of contemporary knowledge societies.

Finally, there is the remarkable resilience of advanced capitalist democracies confronted with major "shocks" from their beginnings in the early 20th century through one of the most turbulent modern centuries. Democracy is a more effective adaptive organism than other forms of governance (Luce, 2017:87). Although the past is not necessarily a solid foundation for anticipating the future state of affairs, there is "a near-zero probability of rich democracies reverting to authoritarianism" (Iversen and Soskice, 2019; see also Przeworski and Limongi, 1997). Obviously, there are exceptions. We just do not know yet if the exceptional circumstances of climate change in the future present will be of such magnitude that the past indeed is not any guide to the future health of democracies.

4.5 What is to be done? Enhancing democracy?

What is good governance under exceptional circumstances? Is democratic governance effective governance? And why should a more democratic as well as egalitarian society be beneficial as the socio-political foundation for coping with extreme circumstances?

The discourse of the impatient scientists in their disenchantment with democracy privileges hegemonic players such as world powers, central states, transnational organizations, and multinational corporations. It is argued that societies with a top-down, centralized socio-political control are better able to confront a major crisis such as the pandemic or climate change (cf. Tooze, 2021; Maçães, 2021; Krugman, 2022). Participatory strategies are only rarely in evidence. Likewise, global mitigation has precedence over local adaptation. "Global" knowledge triumphs over "local" knowledge. However, societal trends appear to operate into the opposite direction. The ability of large societal institutions to impose their will on citizens is declining (Stehr, 2001). As a result, people mobilize around local concerns and efforts including those of the consequences of climate change — thereby enhancing the democratic in democratic governance.

The discussion of options for future climate policies supports the impression that the same failed climate policies must remain in place and are the only correct approach; it is simply that these policies have to become more effective and "rational." It follows that international negotiations must lead to an agreement for concrete, but much broader emission reduction targets. Only a super-Kyoto can still help us. But how the noble goals of a comprehensive emission reduction can be practically and politically *enforced* remains in the fog of general declarations of intent and only sharpens the political skepticism of scientists.

The still dominant line of attack to climate policy shows little evidence of success neither at the state level nor at the global scale. On the contrary, everything that continues to be set in motion worldwide aims at a persistent economic growth which prevents those emissions declining. An alternative model is needed: A model in which action under ambivalent, uncertain and unexpected circumstances can be compelled. A model that recognizes moreover that climate change is a wicked problem that can only be attacked indirectly and requires persistence over a longer period of time. That kind of model will only be found through revitalized rather than less democratic interaction.

Climate policy must be compatible with democracy otherwise the threat to civilization will be much more than just changes to our physical environment. Climate change demands for complex solutions that require worldwide empowerment and knowledgeability of individuals, groups and movements that labor on environmental issues. More democracy combined with political efforts to move toward a more equitable society could be the key toward sustainable climate policies. More democracy comes by definition with greater political participation especially among those that now typically standing on the sidelines of political participation, for example, the young and the economically disadvantaged strata.[22]

A more egalitarian society "would not necessarily maintain rational ecological policies, but it would be more likely to do so" (Best and Connolly, 1975:59). When life chances are more equally distributed, assuring that no one can escape the benefits and costs of a resolution of a serious public problem,[23] one should expect that

[22] Concrete advice on how to avoid oligarchic tendencies in organization may be found, for example, in Robert K. Merton's (1966) essay on "Dilemmas of democracies in the voluntary association."

[23] The systematic reduction of patterns of social inequality in modern societies enhances democratic governance and political participation (Soci, Maccagnan and Mantovanis, 2014:6).

"the political system is very likely to generate collective responses to common dangers and burdens" (Best and Connolly, 1975:59). The English political scientist David Runciman (2013a:316) spells out two further distinct, practical advantages of democracies over authoritarian governments faced by extraordinary circumstances: "The first is their ability to pull together when the threat becomes too big to ignore [...]. The second is their ability to keep experimenting and adapting to the challenges they encounter."

A war-like footing, in contrast, has exactly the opposite effect. A war-like approach reduces the complexity of social and political life in as much as war "nationalizes people's life. Private activities [... are] largely shaped by collective constraints" (Rosanvallon, [2011] 2013:183) as would be the case under authoritarian rule. Under modern conditions, especially the heightened cognitive and social abilities of ordinary citizens require their political participation for successful policies and good governance.[24]

Moreover, a further denationalization of governance will assist in producing new, multiple forms of social solidarity and obligations, strengthens local/regional responses to climate change and enhances the understanding of social interdependence. In addition, self-sufficiency of social institutions has to be guaranteed and — if necessary — re-created in order to transcend boundaries, joining allegedly distinctive motives and practices of different social institutions, for example, joining economic and moral incentives and enhancing the complexity of needs.

The tendency to overestimate and overreach in assigning a crucial role to the singularity of knowledge (and information) in social conduct is evident as one considers the question of *how much knowledge* is needed to carry a specific task, let alone how deeply and subtly one needs to know it. Curiosity about the question of how much we need to know also extends to the question on what we do not need to know. In the first instance, this happens to be an issue that is rarely systematically examined. Second, the inclination is prevalent to assume that the resource of knowledge is somehow sufficient to carry a specific transaction. A more adequate conjecture would be to expect that most decisions and actions are carried out with rather limited knowledge and information (cf. Akerlof, 1970; Smith, 2015) about future conditions of action and those actors are cognizant of how little knowledge they are typically able to mobilize in many situations. The pressure to act that characterizes everyday life ensures that, despite the limited knowledge and information of most actors, decisions are taken and action taken. That we are often forced to act with limited knowledge is not a constitutive deficiency of democracy.

[24] Hans Jonas ([1979] 1984:146) advances a similar observation about systematic inability of authoritarian governments to transcend policy mistakes.

"Life cannot wait" (Durkheim, [1912] 1965:479; see also Gehlen, [1940] 1988:296–297). In most social contexts, the need to act takes precedence over the need to know.

The erosion of democracy might seem to be "convenient" to some, for example among populists but surely is an unnecessary suppression of social complexity. Friedrich Hayek (1960:25) pointed at a paradoxical development. As science advances, it tends to strengthen the observation that we should "aim at more *deliberate and comprehensive control of all human activities*." Hayek pessimistically adds, "It is for this reason that those intoxicated by the advance of knowledge so often become the enemies of freedom."

That democratic governance is *slow* compared for example, to the speed at which decisions are made in the modern economy (see Stehr and Voss, 2019) cannot be denied. In the eyes of many citizens including of course climate scientists, the slowness and the deliberateness of decision-making generate permanent discontent. Climate scientists with their escalating warnings about imminent risks and dangers of climate change repercussions and their communication about failure of politics to heed these forewarnings do not reduce such restlessness among citizens. It is therefore representing a major challenge for democracies to speed up political decision making as well as enhance opportunities for participation in democratic decision making, for example, in the workplace (cf. Herzog, 2019) and the local political community.

4.6 Conclusions

> *Certain kinds of states, driven by utopian plans and an authoritarian disregard for the values, desires and objections of their subjects, are indeed a mortal threat to human well-being.*
>
> *James Scott (1998:7)*

As *Nature* (December 4, 2014:8) editorializes: "The magnitude of […] climate change is worryingly uncertain. Even more uncertain are the physical, social and economic side effects of global warming. There is every reason to believe that, by and large, they will be harmful." The central issue is no longer whether climate change occurs. It is rather what should be done about it. Climate change is the biggest threat humanity has faced in historical times. Suspending democratic debate and decision-making including extensive citizen participation in order to do what is necessary would either demand elevating experts to become decision makers or delegate power to policymakers (who happen to believe a certain group of experts). Neither the first, the technocratic or social engineering vision, nor the idea of a more authoritarian environmentalism has appeal.

I have collected and advanced arguments that speak to the need to enhance rather than abolish democracy as the best political foundation for policies that address climate change as a wicked problem. It is important to push back against simplified solutions to climate change. In debating, researching and understanding climate and climate change we would do well to heed the complex interconnections of the climate system, but also the societal processes, practices and tensions through which science, society, nature and climate permeate, accompany, cover and envelop each other (for such a theoretical perspective, see Stehr and Machin, 2019).

References

Adam, David (2009) "Leading climate scientist: Democratic process isn't working," *The Guardian*. Available at: http://www.theguardian.com/science/2009/mar/18/nasa-climate-change-james-hansen.

Aitken, Mhairi (2012) "Changing climate, changing democracy: A cautionary tale," *Environmental Politics* 21: 211–229.

Akerlof, George A. (1970) "The market for 'lemons': Quality, uncertainty, and the market mechanism," *The Quarterly Journal of Economics* 84: 488–500.

Beeson, Mark (2010) "The coming of environmental authoritarianism," *Environmental Politics* 19: 276–294.

Best, Jacqueline (2018) "Technocratic exceptionalism: Monetary policy and the fear of democracy," *International Political Sociology* 12: 328–345.

Best, Michael H. and William E. Connolly (1975) "Market images and corporate power: Beyond the 'economics of environmental management'", in Kenneth M. Dolbeare (ed.), *Sage Yearbooks Public Policy Evaluation*. Beverly Hills: Sage, pp. 41–74.

Blumenberg, Hans (2006) *Beschreibungen des Menschen*. Frankfurt am Main: Suhrkamp Verlag.

Bomberg, Elizabeth (2017) "Environmental politics in the Trump era: An early assessment," *Environmental Politics* 10: 1–8.

Bozeman, Barry and Juan D. Rogers (2002) "A churn model of scientific knowledge value: Internet researchers as a knowledge value collective," *Research Policy* 31: 769–794.

Churchman, C. West (1967) "Wicked problems," *Management Science* 14: B141–B142.

Clark, Nigel and Kathryn Yusoff (2017) "Geosocial formations and the anthropocene," *Theory Culture & Society* 34: 3–23.

Crouch, Colin (2004) *Post-Democracy*. Cambridge: Cambridge University Press.

Di Paola, Marcello and Dale Jamieson (2018) "Climate change and the challenges to democracy," *University of Miami Law Review* 72: 369–424.

Downs, Anthony (1972) "Up and down with ecology — The issue-attention cycle," *Public Interest* 28: 38–50.

Durkheim, Emile ([1955] 1983) *Pragmatism and Sociology*. Cambridge: Cambridge University Press.

Durkheim, Emile ([1912] 1965) *The Elementary Forms of Religious Life*. New York: Free Press.

Elias, Norbert (1984) "Knowledge and power," pp. 251–292 in Nico Stehr and Volker Meja (eds), *Society and Knowledge. Contemporary Perspectives on the Sociology of Knowledge*. New Brunswick, New Jersey: Transaction Books.

Foucault, Michel ([1981–1985] 2005) „Die Maschen der Macht," pp. 224–244 in Michel Foucault (ed.), *Schriften*. Volume 4. Frankfurt am Main: Suhrkamp.

Fischer, Frank (2017) *Climate Crisis and the Democratic Prospect. Participatory Governance in Sustainable Communities*. Oxford: Oxford University Press.

Fukuyama, Francis (2018) *Identity. Contemporary Identity Politics and the Struggle for Recognition*. London: Profile Books.

Gauchat, Gordon and Kenneth T. Andrews (2018) "The cultural-cognitive mapping of scientific professions," *American Sociological Review* 83: 567–595.

Gehlen, Arnold ([1940] 1988) *Man. His Nature and Place in the World*. New York: Columbia University Press.

Giddens, Anthony (2009) *The Politics of Climate Change*. Cambridge: Polity Press.

Gigerenzer, Gerd and Rocio Garcia-Rettamero (2017) "Cassandra's regret: The psychology of not wanting to know," *Psychological Review* 124: 179–196.

Gilley, Bruce (2012) "Authoritarian environmentalism and China's response to climate change," *Environmental Politics* 21: 287–307.

Granovetter, Mark (1985) "Economic action and social structure: The problem of embeddedness," *American Journal of Sociology* 91: 481–510.

Grundmann, Reiner and Nico Stehr (2012) *The Power of Scientific Knowledge. From Research to Public Policy*. Cambridge: Cambridge University Press.

Hansen, James (2009) *Storms of My Grandchildren*. London: Bloomsbury.

Haldane, Andrew ([2009] 2013) "Rethinking the financial network," in Stephan Jansen, Eckhard Schröter and Nico Stehr (eds.), *Fragile Stabilität — stabile Fragilität*. Wiesbaden: Springer VS, pp. 243–278.

Hayek, Friedrich August von (1944) *The Road to Serfdom*. London: George Routledge & Sons.

Hayek, Friedrich August von (1960) *The Constitution of Liberty*. London: Routledge.

Herzog, Lisa (2019) *Die Rettung der Arbeit*. München: Hanser Berlin.

Hobsbawn, Eric (2007) *Globalisation, Democracy and Terrorism*. London: Abacus.

Iverson, Torben and David Soskice (2019) *Democracy and Prosperity: Reinventing Capitalism through a Turbulent Century*. Princeton University Press.

Jasanoff, Sheila (2012) *Science and Public Reason*. London and New York: Routledge.

Jamieson, Dale (2014) *Reason in a Dark Time. Why the Struggle against Climate Change Failed — and What It Means for Our Future*. New York: Oxford University Press.

Jonas, Hans ([1979] 1984) *The Imperative of Responsibility. In Search of an Ethics for the Technological Age*. Chicago and London: University of Chicago Press.

Kahneman, Daniel (2003) "Maps of bounded rationality: A perspective on intuitive judgment and choice," pp. 449–489 in Tore Frängsmyr (ed.), *Les Prix Nobel: The Nobel Prizes 2002*. Stockholm: Nobel Foundation.

Keller, Evelyn Fox (2017) "Climate science, truth and democracy," *Studies in History and Philosophy of Biological and Biomedical Sciences* 64: 106–122.

Kennel, Charles F. (2013) "Speaking scientific truth to power," *The Cambridge Journal of Anthropology* 31: 150–155.

Krugman, Paul (2022) "Covid's economic mutations," *New York Review of Books* 69 (March 10): 19–20.

Kuklinski, James D. (1990) "Information and the study of politics," pp. 391–395 in John A. Forejohn and James H. Kuklinski (eds), *Information and the Democratic Processes*. Urbana and Chicago: University of Illinois Press.

Levitsky, Steven and Daniel Ziblatt (2018) *How Democracies Die. What History Reveals about the Future*. London: Viking.

Lipset, Seymour Martin, Martin Trow, and James S. Coleman ([1956] 1962) *Union Democracy: The Internal Politics of the International Typographical Union*. New York: Doubleday & Company.

Lovelock, James (2009) *The Vanishing Face of Gaia*. New York: Basic Books.

Lowe, Adolph (1971) "Is present-day higher education learning 'relevant'?" *Social Research* 38: 563–580.

Luce, Edward (2017) *The Retreat of Western Liberalism*. London: Abacus.

Luhmann, Niklas (1976) "The future cannot begin: Temporal structure in modern society," *Social Research* 43: 130–152.

Luhmann, Niklas ([1986] 1989) *Ecological Communication*. Chicago: University of Chicago Press.

Luhmann, Niklas ([1992] 1998) *Observations on Modernity*. Stanford, California: Stanford University Press.

Luhmann, Niklas (2005) *Risk. A Sociological Theory*. With a New Introduction by Nico Stehr and Gotthard Bechmann. London: Aldine Transaction.

Maçães, Bruno (2021) *Geopolitics for the End Time*. London: Hurst Publishers.

McKibben, Bill (2016) "We're under attack by a powerful enemy — and our only hope it to mobilize like we did in WWII," *New Republic* (September): 22–321.

McKibben, Bill (2018) "A Very Grim Forecast," *New York Review of Books* (October 25, 2018) https://www.nybooks.com/articles/2018/11/22/global-warming-very-grim-forecast/.

Marquand, David (2004) *The Decline of the Public. The Hollowing Out of Citizenship*. Cambridge: Polity Press.

Marshall, George (2014) *Don't Even Think About It. Why Our Brains Are Wired to Ignore Climate Change*. New York: Bloomsbury.

McDonald, Susan (2009) "Changing climate, changing minds," *International Journal of Sustainable Communities* 4: 45–63.

Merton, Robert K. (1966) "Dilemmas of democracy in the voluntary associations," *American Journal of Nursing* 66: 1055–1061.

Michels, Robert ([1915] 1949) *Political Parties: A Sociological Study of the Oligarchical Tendencies of Modern Democracy*. New York: Free Press.

Monastersky, Richard (2015) "Anthropocene: The human age," *Nature* 519: 144–147.

Mounk, Yascha (2018) *The People vs. Democracy. Why Our Freedom Is in Danger & How to Save It*. Cambridge, Massachusetts: Harvard University Press.

Nordhaus, William (2015) *The Climate Casino. Risk, Uncertainty, and Economics for a Warming World*. New Haven, Connecticut: Yale University Press.

Petersen, Michael Bang and Lene Aarøe (2013) "Politics in the mind's eye: Imagination as a link between social and political cognition," *American Political Science Review* 107: 275–293.

Pielke, Roger Jr. (2010) *The Climate Fix. What Scientists and Politicians Won't Tell You About Global Warming*. New York: Basic Books.

Peters, B. Guy (2017) "What is so wicked about wicked problems? A conceptual analysis and a research program," *Policy and Society* 36(3): 385–396.

Piketty, Thomas ([2013] 2014) *Capital in the Twentieth Century*. Cambridge, Massachusetts: Harvard University Press.

Prewitt, Kenneth (2010) "Introduction: Limits to knowledge? No easy answer," *Social Research* 77: 901–904.

Prins, Gwyn, Isabel Galiana, Professor Christopher Green, Reiner Grundmann, Mike Hulme, Atte Korhola, Frank Laird, Ted Nordhaus, Roger Pielke Jr., Steve Rayner, Daniel Sarewitz, Michael Shellenberger, Nico Stehr, and Hiroyuki Tezuka (2010) *Hartwell Paper I*. London: London School of Economics.

Przeworski, Adam and Fernando Limongi (1997) "Modernization: Theories and facts," *World Politics* 49: 155–183.

Rosanvallon, Pierre ([2011] 2013) *The Society of Equals*. Cambridge, Massachusetts: Harvard University Press.

Rosanvallon, Pierre (2006) *Democracy Past and Future*. New York: Columbia University Press.

Rhodes, Ekaterina, Jonn Axsen, and Mark Jaccard (2014) "Does effective climate policy require well-informed citizen support?" *Global Environmental Change* 29: 92–104.

Ringen, Stein (2016) *The Perfect Dictatorship: China in the 21st Century*. Hongkong: Hongkong University Press.

Rittel, Horst and Melvin M. Webber (1973) "Dilemmas in the general theory of planning," *Policy Sciences* 4: 154–159.

Rorty, Richard (2004) "Post-democracy," *London Review of Books* 26: 10–11.

Runciman, David (2018) *How Democracy Ends*. London: Profile Books.
Runciman, David (2013a) *The Confidence Trap. A History of Democracy in Crisis from World War I to the Present*. Princeton, New Jersey: Princeton University Press.
Runciman, David (2013b) "Democracy's dual dangers," *The Chronicle of Higher Education*. November 18. https://www.chronicle.com/article/Democracys-Dual-Dangers/142971.
Rykkja, Lise, Simon Neby, and Kriston L. Hope (2014) "Implementation and governance: Current and future research on climate change policies," *Public Policy and Administration* 29: 106–130.
Sarewitz, Daniel (2010) "Normal science and the limits on knowledge: What we seek to know, what we choose not to know, what we don't bother knowing," *Social Research* 77: 997–1010.
Schon, David A. ([1963] 1967) *Invention and the Evolution of Ideas*. London: Tavistock.
Schumpeter, Joseph A. (1942) *Capitalism, Socialism and Democracy*. New York: Harper-Collins.
Scott, James C. (1998) *Seeing like a State. How Certain Schemes to Improve the Human Condition Have Failed*. New Haven, Connecticut: Yale University Press.
Seefried, Elke (2015) *Zukünfte. Aufstieg und Krise der Zukunftsforschung*. Berlin: de Gruyter.
Shearman, David and Joseph Wayne Smith (2007) *The Climate Change Challenge and the Failure of Democracy*. London: Praeger.
Simmel, Georg (1890) *Über sociale Differenzierung: Sociologische und psychologische Untersuchungen*. Leipzig: Duncker & Humblot.
Simmel, Georg ([1907] 1989) *Philosophie des Geldes. Gesamtausgabe Band 6*. Frankfurt am Main: Suhrkamp.
Skidelsky, Robert and Edward Skidelsky (2012) *How much is Enough? Money and the Good Life*. New York: Other Press.
Soci, Anna, Anna Maccagnan, and Daniela Mantovani (2014) "Does inequality harm democracy? An empirical investigation on the UK," *6th International Scientific Conference on Economic and Social Development and 3rd Eastern European ESD Conference: Business Continuity*, Vienna, 24–25 April 2014.
Smith, Charles (2015) *What the Market Teaches Us. Limitations of Knowing and Tactics for Doing*. Oxford: Oxford University Press.
Stehr, Nico (1991) "The power of scientific knowledge — and its limits," *Canadian Review of Sociology and Anthropology* 29: 460–482.
Stehr, Nico (1994) *Knowledge Societies*. London: Sage.
Stehr, Nico (1997) "Trust and climate," *Climate Research* 8: 163–169, 199.
Stehr, Nico (2001) *The Fragility of Modern Societies: Knowledge and Risk in the Information Age*. London: Sage.
Stehr, Nico (2015) "Democracy is not an inconvenience," *Nature* 525: 449–450.
Stehr, Nico (2016a) *Information, Power, and Democracy: Liberty Is a Daughter of Knowledge*. Cambridge: Cambridge University Press.

Stehr, Nico (2016b) "Exceptional circumstances. Does climate change trump democracy?" *Issues in Science and Technology* 32: 37–44.

Stehr, Nico and Amanda Machin (2016a) "Inequality in modern society: Causes, consequences," in Nico Stehr and Amanda Machin (eds), *Understanding Inequality: Social Costs and Benefits*. Wiesbaden: Springer VS, pp. 3–36.

Stehr, Nico and Amanda Machin (2016b) "Trusting the climate: Catastrophe vs. stability," *Society* 53: 573–580.

Stehr, Nico and Amanda Machin (2019) *Society & Climate. Transformations and Challenges*. Singapore: World Scientific Publishers.

Stehr, Nico and Dustin Voss (2019) *Money. A Social Theory of Modernity*. New York: Routledge.

Stehr, Nico and Marion Adolf (2018) *Ist Wissen Macht? Wissen als gesellschaftliche Tatsache*. Weilerswist: Velbrück Wissenschaft.

Swyngedouw, Erik (2010) "Apocalypse forever? Post-political populism and the spectre of climate change," *Theory, Culture and Society* 27: 213–232.

Swyngedouw, Erik (2011) "Interrogating post-democracy: Reclaiming egalitarian political spaces," *Political Geography* 30: 370–380.

Tenbruck, Friedrich H. (1977) "Grenzen der staatlichen Planung," in Wilhelm Hennis, Peter Graf Kielmansegg, and Ulrich Matz (eds.), *Regierbarkeit. Studien zu ihrer Problematisierung. Band 1*. Stuttgart: Klett-Cotta, pp. 134–149.

Tooze, Adam (2021) *Shutdown: How Covid Shook the World's Economy*. London: Penguin.

Tsing, Anna (2015) *The Mushroom at the End of the World: On the Possibility of Life in Capitalist Ruins*. Princeton, New Jersey: Princeton University Press.

USGCRP (2018) *Impacts, Risks, and Adaptation in the United States: Fourth National Climate Assessment, Volume II* [Reidmiller, D.R., C.W. Avery, D.R. Easterling, K.E. Kunkel, K.L.M. Lewis, T.K. Maycock, and B.C. Stewart (eds.)]. U.S. Global Change Research Program, Washington, DC, USA. doi: 10.7930/NCA4.2018.

Van Dijk, Teun A. (2014) *Discourse and Knowledge. A Sociocognitive Approach*. Cambridge: Cambridge University Press.

Venn, Couze (2018) *After Capital*. London: Sage.

Weber, Max ([1918] 1994) "Parliament and government in Germany under a new political order," in Peter Lassman and Ronald Spiers (eds.), *Weber. Political Writings*. Cambridge: Cambridge University Press, pp. 130–271.

Williamson, Oliver E. (1975) *Markets and Hierarchies. Analysis and Antitrust Implications*. New York: Free Press.

Williamson, Oliver E. (1985) *The Economic Institutions of Capitalism. Firms, Markets, Relational Contracting*. New York: Free Press.

Williamson, Oliver E. (1995) "Transaction cost economics and organization theory," pp. 77–107 in Neil J. Smelser and Richard Swedberg (eds.), *The Handbook of Economic Sociology*. Princeton, New Jersey: Princeton University Press.

Wissenschaftlicher Beirat der Bundesregierung „Globale Umweltveränderungen" (WBGU) (2012) *Welt im Wandel — Gesellschaftsvertrag für eine Große Transformation*. Berlin: Wissenschaftliche Beirat der Bundesregierung „Globale Umweltveränderungen."

Zakaria, Fareed ([2004] 2007) *The Future of Freedom. Illiberal Democracy at Home and Abroad*. New York: W.W. Norton.

5. A Very Blind Spot¶

The need for action in the context of the climate crisis cannot be more urgent. This is why the recent heated climate debate in many countries has its good points. The climate issue is once again at the top of the urgent political agenda in Europe and elsewhere. Although there is fierce debate about politically and socially enforceable solutions, it is almost always and still only one-sided — with proclamatory demands for symbolic measures, ignoring the global source of global warming and the need for social adaptation to climate change.

The political and societal debate in the United States and in Europe continues to focus on measures to reduce national and/or global emissions of greenhouse gases. These include not only pricing CO_2 emissions, but also afforestation of our tree population, the promotion of renewable energies — for example, the construction of wind turbines, electric mobility, insulation of real estate, and the reduction of value-added tax on rail tickets, etc.

The available political capital is invested exclusively in avoiding national emissions.

These are undoubtedly undertakings worthy of promotion, but with very limited effectiveness in tackling the problem of greenhouse gases accumulating in the atmosphere and their effect on the climate. In addition, there is the opinion of many that they can use the good opportunity to put other issues on the agenda that have a significant impact on the climate — such as the issue of air quality and the use of diesel vehicles in cities or ship exhaust gases in ports, speed limits on motorways, animal welfare or plastic in the sea.

A phenomenon that is little discussed in public, however, is the length of time greenhouse gases remain in the atmosphere. The retention time of different gases varies. It takes centuries for the additional gases to leave the atmosphere. We do not have precise findings, or, to put it another way, the reversibility of man-made climate change is an uncertain quantity. A recent study estimates that climate change would only reverse 1000 years after emissions have been completely halted. In other words, anthropogenic climate change is irreversible for at least a millennium.

So the climate changes that have occurred so far and will materialize in the years to come will continue and eventually be seen as "normal," even if the ambitious plan to stop the release of greenhouse gases is successful. This means that climate change is there, we can limit it, but we have to live with climate change. Mitigation reduces these changes, but it does not undo them. Politicians, society and science should urgently address not only mitigation but also precautionary

¶This section originally appeared in Stehr, N. and H. von Storch: "A very blind spot," *Society* 56: 611–612, 2019.

measures in response to the consequences of climate change. This is complicated by three factors:

1. There are no coordinated time scales for sustainable moderation and adaptation outcomes. The successes of moderating greenhouse gas emissions only become apparent in the distant future. Even the immediate implementation of lower CO_2 emissions does not come in time to limit climate change in the coming few decades. As long as greenhouse gases are released anywhere in the world, the climate will continue to change. The unlimited emissions to date ensure that climate change will change the way we live.
2. The threat posed by climate-induced extreme events, such as heavy rainfall, flooding, and heat waves, is already very high in many regions of the world. Just think of New Orleans. The vulnerability of our livelihoods increases to the extent that the growing world population settles in regions that are endangered, where growing population groups are marginalized without protection and, due to the political economy, become victims of so-called natural disasters.
3. The regions of the world whose livelihoods will be particularly affected by the consequences of global climate change, in particular the least developed countries, are already rightly and increasingly demanding that the world should concern itself with their protection and not only with the protection of the climate.

Despite the hitherto contrary practice of all political parties to speak of climate protection programs, adaptation as a precautionary measure is politically much easier to implement and legitimize than mitigation strategies; it is also attractive because its success will not occur in the distant future. When it comes to finding solutions to a problem through innovations in science and technology, for example, these can be presented much better if they are conceived as adaptation measures. Adaptation strategies also make it easier to achieve several goals at once: Improving quality of life, air quality, medical care, reducing social inequality and increasing political participation are not mutually exclusive.

Adaptation processes can become the engine of what we call sustainable economic activity today. Adaptation can lead to a reduction in greenhouse gas emissions; adaptation and moderation do not contradict each other.

However, reducing emissions alone does not necessarily lead to adaptation. All sustainability is local. It is not just a question of raising coastal dikes, but of a bundle of measures in the health sector, in mobility, expectations of living space, water supply, land use, socialization patterns, democracy or the management of coastal ecosystems. In the coming decades, we will have to think increasingly about what is feasible. And an essential part of what is feasible is precaution — to the benefit of all of us.

To put it briefly and radically, we should start living with the inevitable climate change and its challenges. Private and public funds are needed to enable problem-based preventive research for all areas of human life. So far, this issue has not really arrived in the public debate. Although the technical departments of companies and administrations have long since been thinking about adapting to changes to be expected in the future, it seems that business and politics are still terrified to pronounce the word adaptation or precaution.

That must change. Adaptation measures and mitigation measures are both significant political goals. The risks and dangers of dealing with the practical consequences of climate change in the form of adaptation should be high on the political agenda.

6. Adaptation versus Mitigation. On the Politics of Terminology in the Climate Discussion[‖]

Hansvolker Ziegler (2008) in his polemic — a polemic because his contribution is full of contradictions and stubbornly interprets other positions — praised the virtues of the supposedly prevailing conceptualization, namely that of the *International Panel on Climate Change* (IPCC). Ziegler argues that the IPCC has always recognized and demanded *adaptation and mitigation* as necessarily interrelated approaches to climate policy. He accuses us of denying this by constructing a nonsensical dichotomy between adaptation and mitigation, thereby doing a disservice to science, policy, and the goal of sustainability.

There are those, as expressed in Ziegler's concerned tenor, who have "recently" opened up unfamiliar fronts in the debate over the supposedly converging political strategies in response to global warming. They are, in politically correct terms, "so-called climate skeptics" and "climate deniers" who are not only inspired by the illusion of the difference between adaptation and avoidance but who have also successfully infected "some scientific quarters" with their abstruse ideas. This circle of people, whose size is not revealed to us, has still not "committed itself to sustainability."

Two aspects should be elaborated in Ziegler's polemic: First, the intellectual impurity according to which adaptation and avoidance are always and necessarily two equal sides of the same coin, which for strange reasons we artificially and confusingly present to the public as socially and scientifically separate approaches; second, the latent request of a representative of the political administration to science to keep in mind the usability of science for politics. We limit ourselves here to the first aspect, a discussion of the balance (convergence) and conditionality (difference) of adaptation and avoidance.

To avoid misunderstandings, let us clarify that mitigation or "avoidance" (also "reduction," "moderation") is about reducing or eliminating the human causes of warming and its consequences, especially emissions of greenhouse gases such as carbon dioxide or methane. Adaptation is about enabling society to cope with climate hazards, especially those that will become more severe in the future, by acting before damage occurs.

However, if damage is already occurring, adaptation is about mitigating damage. Of course, adaptation and mitigation may be linked: On the one hand, forests

[‖]This section originally appeared in Stehr, N. and H. von Storch: "Anpassung und Vermeidung oder von der Illusion der Differenz" [Adaptation and mitigation or the illusion of a difference], 2008. Response to: Ziegler, H.: "Adaptation versus mitigation Zur Begriffspolitik in der Klimadebatte", *GAIA* 17(1): 19–24, 2008 [translated by the authors].

can serve to store carbon, but on the other hand, they can also provide slope stability or more favorable conditions for the storage of precipitation, which would otherwise flow unchecked to the rivers. As a rule, however, measures serve either prevention or adaptation.

In fact, we do not consider adaptation and mitigation to be alternatives, as Ziegler implies to us, but necessary parts of an overall strategy. Moreover, mitigation is useful when adaptation largely succeeds, and adaptation is useful when mitigation largely succeeds. When we emphasize adaptation in our work, this does not mean that we are critical of mitigation strategies, but only of a systematic neglect of adaptation measures by science and politics.

6.1 Of the virtue of one-sidedness

Not only scientific efforts but also climate policy is overwhelmingly one-sided when it comes to the possible difference or convergence of mitigation and adaptation. Ziegler denies that there can or may be such a difference, but his argument thrives on the fact that this is precisely how the distinction is made in certain quarters of science. Rather, what is correct — and we have argued this position for many years — is that there is an almost singular concentration on avoidance strategies in publicly visible and effective science and in politics. How one can overlook the fact of differentiation between adaptation and mitigation in the practice of science and policy to date is puzzling to us.

Ziegler in concert with parts of the scientific community and especially climate policy seems to tacitly assume that successful mitigation makes an adaptation strategy unnecessary. This is precisely what makes the seriousness of such a climate strategy implausible. Ziegler refers to an IPCC consensus of 2007 (Ziegler, 2008, footnote 2), according to which the combination of adaptation and mitigation could significantly reduce many (exactly Which? Where? When?) risks of climate change. Nevertheless, the IPCC subsequently emphasizes: "Many (again: Exactly Which? Where? When?) impacts (of climate change) can be reduced, delayed or avoided by mitigation" (quoted from Ziegler, 2008, footnote 2).

At the same time, Ziegler (2008, p. 20) claims — without basing this on evidence, as far as can be seen — that "strategies to stabilize and reduce greenhouse gases (GHG) that take effect as soon as possible (are) more effective and less costly in the long run, because they also help to reduce the damage and costs of adapting to climate change that has already occurred or that can no longer be avoided because of the inertia of the systems." So, mitigation as a priority, which is not only less expensive but also (intended or not?) eliminates the need for adaptation measures?

6.2 From the taboo of a difference

We would like to oppose this classical, but contradictory position with our — however distortedly presented by Ziegler — considerations. These theses, which are not at home in the scientific quarters of the deniers of anthropogenic climate change, can be said to be both realistic, i.e., based on solid scientific findings, and, following their realism, to assume a convergence of adaptation and mitigation as well as the difference between corresponding research priorities and climate policy. In the following, we try to substantiate our position in nine theses.

In light of Ziegler's polemic, we focus on why it is by no means "nonsensical" or a "bogus alternative" to make a distinction between adaptation and mitigation, leaving aside the fact that the two can converge or become indistinguishable to a limited extent. Ziegler's position can be illustrated with a homeowner who lives in a low-energy house (mitigation) and therefore feels safe from climate hazards. He forgets that one day his house could be up to the roof in water, his roof could fly off, slide down the slope, or become uninhabitable in an extremely hot and dry period. He neglects precautionary adaptation in association with the prevailing climate change policy.

1. Global warming is not a temporary or short-lived phenomenon. This statement is relevant because often — consciously or not — the impression is given that the climate can be changed within a short period of time. At the same time, uncertainty (in the sense of Knight, 1921) is one of the fundamental characteristics of any analysis of the forward-looking climate problem. It is not the fundamental mechanism of warming that is uncertain, but its natural and social consequences. We live in a fragile world (see Stehr, 2000), in which probability distributions of the consequences of climate change are not (yet) available.

   Ending global warming as defined by the United Nations Framework Convention on Climate Change requires a reduction of anthropogenic greenhouse gas emissions to near zero, which is only possible with tremendous global efforts. Beyond that, it will take several decades to centuries for increased CO_2 concentrations to return to preindustrial equilibrium. Even if it were possible to reduce emissions by 80 percent in just one year, the climate would not reach a new equilibrium for decades. In other words, the climate change underway cannot be stopped overnight, no matter how great the efforts of mitigation policy. A climate policy that is predominantly dedicated to the problem of mitigation, ignoring the pressure to adapt, is therefore irresponsible. The goal of such a policy, to protect the climate from society — and thus society from itself — can only be achieved in the distant future.

2. Both global and German climate policy is governed by the Kyoto Protocol. This deals almost exclusively with mitigation issues. The reduction targets of the Kyoto Protocol, which expires in 2012, are unlikely to be achieved. At best, they would reduce the temperature increase anticipated by 2012 by 0.1 degree Celsius. The so-called clean development mechanism of the Kyoto Protocol would delay the amount of global cumulative emissions by one week until 2012 compared to a situation without Kyoto reductions.

   There is no obligation for developing countries and emerging economies, especially China and India, to reduce greenhouse gas emissions. We do not have precise data on the greenhouse gas emissions of these countries, but we can assume that their share of global emissions is steadily increasing. Emissions from industrialized countries are also likely to (continue to) increase despite all mitigation efforts. The Kyoto approach as a socially restrictive, large-scale global planning has failed. A successor process based on this hegemonic planning mentality will not be effective (cf. Scott, 1998; Prins and Rayner, 2007).
3. As a result, climate change is progressing steadily and will move up a gear in the future. Reversing the change in our earth's climate is only possible in decades, if not centuries.
4. There are at least three important arguments why policymaking, society and science should consider not only protection of climate but also adaptation to the impacts of climate change (cf. Pielke *et al.*, 2007):

   (a) The emissions to date ensure that climate change will change our living conditions. The successes of mitigation, on the other hand, will only become apparent in the distant future. The dilemma is that the time scales of nature are not congruent with those of socio-political decision-making cycles in democratic societies, which are reflected, for example, in election periods and attention cycles, but also in people's fundamental horizons for action.

   (b) The threat posed by weather-related extreme events such as heavy rain, floods, drought, mudslides and heat waves is already considerable in many regions of the world. Just think of New Orleans, Myanmar or Hurricane Mitch, which was instrumentalized during the negotiations in Rio de Janeiro in 1992. The vulnerability of our livelihoods increases as the growth of the world's population takes place in vulnerable regions and as growing populations are marginalized without protection, who then become victims of so-called natural disasters due to political economy. Total security cannot guarantee a purposeful, proactive adaptation policy. But it can mitigate vulnerability to political-economic conditions.

(c) The regions of the world that will be particularly affected by the consequences of climate change are rightly and increasingly demanding that the world take care of their direct protection and not only of climate protection.

5. A telling example of the prevailing one-sidedness of the debate and climate change efforts is the often dispassionately used term "heat deaths." As if people are only victims of nature and not victims of certain social conditions that expose those affected to extreme heat and do not protect them preventively. To speak of heat deaths — as in the summer of 2003 — ultimately only protects the municipalities, regions or states that fail in their duty of precaution. The use of the term virtually guarantees that the developments underlying it will be repeated due to thoughtlessness.
6. Climate change is also an almost perfect example of the tragedy of the commons: The polluters of climate change are hardly asked to pay, although they enjoy the benefits of their actions. Extending this view both in time and space, it will be future generations and the less developed countries that will have to shoulder the consequences of climate change. Precautionary adaptation measures can mitigate these consequences.
7. Despite the seemingly contrary views of all political parties and their reluctance to speak publicly about climate adaptation programs, adaptation as a precautionary measure is politically much easier to implement and to legitimize politically than reducing emissions. Last but not least, it has the advantage that its success does not only occur in the distant future. Adaptation measures can be tailored to the interests of different population groups. Coordination and information deficits can be more easily eliminated. When it comes to finding solutions to a problem through innovations in science and technology, these are easier to present if they are conceived as adaptation measures.
8. Adaptation strategies also make it easier to achieve several goals at once: Improving the quality of life, reducing social inequality, and increasing political participation. Risks and dangers in dealing with uncertainties, such as new technologies, are lower in the case of adaptation measures. Undoubtedly, adaptation measures as such do not automatically achieve these multiple goals; accompanying political and societal framework conditions are also needed.
9. Adaptation processes can become a driver for sustainable economic activity by reducing greenhouse gas emissions (e.g., lower water consumption through efficient dishwashers and thus lower energy consumption), because adaptation and avoidance are not contradictory. The only thing is that avoidance alone will not necessarily lead to adaptation in the coming decades. All

sustainability is local. It is not just a matter of raising coastal dikes, but a bundle of measures in health care, water supply, or marine ecosystem management. In the coming decades, we will increasingly have to think about what is feasible. And what is feasible is careful adaptation — for the benefit of us all.

In short: We should set about surviving. We can therefore only demand that *additional* private and public funds finally be made available for intelligent, comprehensive adaptation research in the social and natural sciences. Of course, this does not mean discarding the previous climate protection goals. Business and politics are afraid to utter the word adaptation because this could be interpreted as giving up, as accepting hubris. And there are those who hide behind the bogus argument that there is no difference between adaptation and mitigation. This must change.

We thank Reiner Grundmann and Hermann Strasser, as well as a GAIA reviewer, for their constructive comments; however, we are solely responsible for the positions taken in this essay.

Literature

Heal, G. 2008. *Climate economics: A meta-review and some suggestions.* NBER Working Paper 13927. www.nber.org/papers/w13927 (retrieved 14.05.2008).

Keenlyside, N. S., M. Latif, J. Jungclaus, L. Kornblueh, E. Roeckner. 2008. Advancing decadal-scale climate prediction in the North Atlantic sector. *Nature* 453: 84–88.

Klinenberg, E. 2002. *Heat wave — A social autopsy of disaster in Chicago.* Chicago, IL: University of Chicago Press.

Knight, F. 1921. *Risk, uncertainty and profit.* Boston, MA: Hart, Schaffner & Marx.

Matthews, H. D., K. Caldeira. 2008. Stabilizing climate requires near-zero emissions. *Geophysical Research Letters* 35: L04705.

Pielke, R. Jr. 2007. *The honest broker: Making sense of science in policy and politics.* Cambridge, UK: Cambridge University Press.

Pielke, R. Jr., G. Prins, S. Rayner, D. Sarewitz. 2007. Lifting the taboo on adaptation. *Nature* 445: 597–598.

Prins, G., S. Rayner. 2007. *The wrong trousers — Radically rethinking climate policy.* Discussion Paper. Oxford, UK: James Martin Institute for Science and Civilisation, Oxford University.

Prisching, M. 2006. *Good Bye New Orleans: Der Hurrikan Katrina und die amerikanische Gesellschaft.* Graz: Leykam.

Scott, J. C. 1998. *Seeing like a state — How certain schemes to improve the human condition have failed.* New Haven, Connecticut: Yale University Press.

Stehr, N. 2000. *Die Zerbrechlichkeit moderner Gesellschaften.* Göttingen: Velbrück.
Stern, N. 2006. *The economics of climate change: The Stern Review.* https://webarchive.nationalarchives.gov.uk/ukgwa/20100407172811/https:/www.hm-treasury.gov.uk/stern_review_report.htm (retrieved 14.05.2008).
Wood, R. 2008. Natural ups and downs. *Nature* 453: 43–44.
Ziegler, H. 2004. Warum nur tut sich die Wissenschaft mit dem Vorsorgeprinzip so schwer? *GAIA* 13(4): 241–247.
Ziegler, H. 2008. *Adaptation versus mitigation* — Zur Begriffspolitik in der Klimadebatte. *GAIA* 17(1): 19–24.

Chapter 6

Outlook

1. *Zeppelin Manifesto* on Climate Protection from 2008*

The policy of climate protection, with the support of influential circles within climate research, is predominantly one-sided. It is not the appropriate way to deal with the problem.

Up to now, it is almost exclusively about measures to do with energy, transport, industry and housekeeping that have been enacted under the heading of climate protection; such as measures to save energy and to increase efficiency, and the corresponding legislative frameworks.

The threat posed to the basic living conditions of society by climatic changes cannot be combated, as it has been up to now, only by protecting the climate from society, particularly given that many of these measures are of a symbolic nature. Additional effective efforts are required on the part of researchers, politicians, and economic leaders in order to come to terms with the climatic dangers that already exist today, and which will intensify in the future, even in the face of a successful climate protection policy. This protection cannot wait to be put in place only after we have lived through catastrophes in the wake of weather extremes; rather, they must be realized in the form of *precautionary* measures. And these are in short supply here and now!

Sometimes such a proposal is countered with the declaration that extending the existing climate protection policy by means of an active precautionary climate

*This section originally appeared Stehr, N. und H. von Storch, 2008: 10-Punkte Manifest: So kann Deutschland den Klimawandel bewältigen [Translated by Paul Malone from a German version]. *Spiegel Online*, http://www.spiegel.de/wissenschaft/natur/0,1518,576032-11,00.html.

policy is essentially identical with admitting that the existing policies have miscarried. This argument is obviously short-sighted and unfounded.

Concentrating climate policy on the reduction of greenhouse gases serves no purpose if it leads at the same time to preventing taking precautions. Such a one-sided research perspective and climate protection policy will neither protect the climate from society in the coming decades nor society from the climate.

In contrast, our *Zeppelin Manifesto* faces up to reality and its demands:

1. Climatic warming is not a fleeting, temporary or short-lived phenomenon. It is important to state this outright, because the impression is often given, intentionally or otherwise, that the climate can be changed in one direction or the other in a short span of time.

   Lowering emissions means, in the first place, only reducing the *increase* in their concentration. And, in fact, it would already be a triumph if we were presently to reduce the increase of these emissions. The long-term prevention of global warming, however, requires a *quite extensive* reduction of greenhouse gas emissions, i.e., lowering human emissions to almost zero. The length of time necessary for our elevated concentration of CO_2 to return even approximately to its original — here, preindustrial — equilibrium amounts to somewhere between several *decades* and a few *centuries*.

   Why are these time spans relevant? On the one hand, they point up the prodigious efforts that are necessary worldwide in order effectively to halt climatic warming; on the other, these numbers are the point of departure for our further theses regarding how society will have to deal with the consequences of climatic warming.
2. Adaptation and prevention, i.e., reduction of emissions, are reasonable options that must be pursued in concert. As a rule, however, they are different options. Adaptation to the dangers posed by the climate will only incidentally reduce emissions; likewise, energy-saving and other reductive measures will only seldom be able to reduce the vulnerability of our basic living conditions in face of the dangers posed by the climate. What both options have in common, however, is that they are promoted by means of technological innovations, but most particularly by means of social changes. A realistic assessment and a public discussion of the dangers of climate change are the first prerequisites for understanding the nature and the extent of the social changes required. A positive atmosphere, in which innovations are actively promoted and publicly acknowledged, is useful not only in the context of an active climate policy.

3. Reductive measures are in any case reasonable and necessary. The same is also true of adaptive measures, which continue to have a lasting effect when the reductive measures begin to work at a later point in time. The more effective the reduction, the more efficacious the adaptive measures — in the long term!
4. Let us proceed, in a thought experiment, from the premise that human beings on this planet could manage to meet the goal of reducing emissions by eighty percent in the space of one year. When, under these conditions, would the climate machine achieve a new "equilibrium"? The answer is: not for decades. In other words, the climatic change that is already underway cannot be prevented overnight, even by the greatest imaginable efforts in the realm of mitigation policy.

   A climate policy that commits itself to the problem of mitigation while neglecting the urgent need for adaptation is an irresponsible climate policy because it denies society's inevitably higher degree of vulnerability in the coming decades. The goal of such a policy — to protect the climate from society, and thereby to protect society from itself — will bear fruit only in the distant future.

   A representative example of the prevailing one-sidedness of the discussion of climate protection and efforts in this area is the often dispassionately employed term "*heat deaths*." As if people were almost inevitably defenseless victims of nature, and not victims of specific social circumstances; and indeed, of social circumstances that irresponsibly put people at the mercy of extreme heat and its consequences, and do not preventively shield the segments of the population that are most severely affected. To speak of "heat deaths," as was done in the case of the hot summer of 2003, protects only the municipalities, regions or countries that failed in their duty to take precautions. The very use of this term guarantees, so to speak, that the trends that are the actual cause of this phenomenon will be thoughtlessly repeated.
5. There are at least three important reasons why politicians, society and scientists must urgently think in terms not only of mitigation but also of *precautionary* measures, as a reaction to the consequences of climate change:

   (a) The time scales of the long-term results of lowering emissions and of climate change do not correspond to each other. Any successes in terms of reducing the emission of greenhouse gases will take effect, as we have said, only in the far future. A world in which only small amounts of CO_2 are still being emitted will come too late to limit climate change in the next decades. The practically unlimited emissions of the past and up to now guarantee that climate change will change our future living conditions. The dilemma lies in the fact that the time scales of nature are not congruent with those of political decision-making cycles in democratic societies, which proceed in terms of election periods and cycles of attention, and which are reflected in the limited horizons of human action.

(b) The threat posed by extreme climatic events, such as torrential rains, floods and heat waves, is already considerable today, and always has been in many regions of the world. One need only recall New Orleans in 2005; the storm surge of 1872 on the German Baltic coast or that of 1953 in Holland; or even Hurricane Mitch, which was turned to good use in the course of the 1992 negotiations in Rio de Janeiro. The vulnerability of our basic living conditions increases parallel to the growth of the global population in endangered regions, where growing segments of the population are marginalized without protection and, not least for reasons of political economy, become victims of extreme weather events.

(c) The regions of the world whose basic living conditions will be particularly hard hit by the consequences of worldwide climatic changes are already demanding today, rightfully and increasingly vehemently, that the world must see to their protection, and not only to the protection of the climate.

6. Worldwide climate policy, like that of Germany as well, is particularly clearly represented by the Kyoto Protocol. The Kyoto process concerns itself almost exclusively with questions of reduction. The reduction targets of the Kyoto Protocol, which expires in 2012, will hardly be achieved. The successful execution of the Kyoto Protocol's so-called "Clean Development Mechanism" (CDM), in terms of the worldwide emission of CO_2, would by 2012 reduce the volume of worldwide cumulative emissions by about a week's worth, compared to the same development without Kyoto reductions.

   For developing and emerging countries, particularly China and India, there is currently no obligation to reduce greenhouse gas emissions. We have no precise data regarding the greenhouse gas emissions produced by these countries, but we can assume that their share of the global balance of greenhouse gases is continually *increasing*. In the future, however, the developed societies will also emit (yet) more climate-damaging greenhouse gases. The total emission of carbon dioxide above all, despite all efforts at reduction, will probably increase further in industrialized countries between now and 2012.

   The Kyoto approach, as a form of socially restrictive, large-scale global planning, has failed. Any subsequent process based on this hegemonic planning mentality will serve no purpose. As a result, climate change of human origin is steadily advancing and will step up in the future. A reversal of this alteration to our global climate will be possible only over the span of decades, if not centuries.

7. Despite the contrary opinions of all political parties up to now and their reluctance to speak publicly about precautionary climate programs, adaptation as a precautionary measure is relatively easy to implement and to legitimize in political terms. Moreover, it has the enormous advantage that its success will be evident in the foreseeable future. When it comes to finding solutions to a problem by means of innovations in science and technology, it is easier to present these in the form of adaptive measures.
8. The consequences of warming vary significantly according to region and climatic zone. Research into precautionary measures thus means expanding our knowledge about regional changes. To what, exactly, are we going to have to adapt? With the aid of adaptive strategies several goals at once can be achieved, because they are primarily locally or regionally oriented, and therefore can be flexibly configured: Improving quality of life, decreasing social inequity and increasing political participation are not mutually exclusive.
9. The dual challenge of adaptation and prevention also leads to a reasonable division of labor. The German federal and European responsibility falls at the level of the frameworks for managing emissions, while for those in charge of the *Länder* and municipalities, the question of reducing their vulnerability should have priority. In fact, institutions and persons charged with specific responsibilities — for coastal protection or for the Hamburg harbor, for instance — demonstrate a concrete commitment to solving problems of adaptation.
10. In the public discussion, down to the present day, prevention alone has been portrayed as a virtuous form of behavior, even when it takes the form of purely symbolic and largely ineffective actions, such as Sundays without driving, doing without long trips, or staging public events. This perception is not unproblematic, to the extent that it gives actors the impression that sufficient steps are being taken to protect the climate. A revision or extension of this perception to include a proactive attitude toward precautions and toward necessary social changes, however, as is essential to protect society from the changing climate and thus to reduce the vulnerability of the very basis of our existence, is still lacking. An effective defense of this basis demands precautionary measures in the coming years and decades. This must now be our priority.

2. Ongoing Related Work by Nico Stehr

We continue to independently examine research questions about the interrelation of climate and society. These include:

1. Climate policy as a "wicked problem": From an analytical point of view, climate policies constitute a "wicked" problem. In contrast to "tame" problems (complicated, but with defined and achievable end-states and simple causal relations), wicked problems comprise open, complex and imperfectly understood systems. Originally described by C. West Churchman (1967) and later explicated more comprehensively by Horst Rittel and Melvin Webber (1973; cf. also Peters, 2017) in the context of urban planning, wicked problems are issues that are often formulated as if they were susceptible to a simple, unilinear solution when in fact they are not. We have examined climate policies in terms of wicked problems properties in series of publications titled the Hartwell Papers: Prins *et al.* (2010, 2013).
2. What is practical *Science in Society: Societies, Climate Change and Policies?* Reflections about the conditions or constituents of practical knowledge have to start from the assumption that the adequacy (usefulness) of knowledge, produced in one context (of production) but employed in another context (of application) pertains to the relation between knowledge and the local conditions of action (Stehr, 1992; Grundmann and Stehr, 2012). Within the context of application constraints and conditions of action are apprehended as either open or beyond the control of relevant actors. Given such a differentiation, practical knowledge pertains to open conditions of action, which means that theoretical knowledge, if it is to be effective in practice, has to be re-attached to the social context in general and those elements of the situation that are actionable in particular.
3. Modern society as a knowledge society: The dominant resource of advanced capitalism are intangible assets (knowledge) and an intangible-intensive production (Stehr, 1994; Stehr, 2001; Stehr, 2015). Compared to the tangible-intensive production of industrial society, the marginal cost of the production of intangibles — as scalable assets — in knowledge societies, i.e., software, standards, organization know-how, platforms and texts, approaches zero. Returns on tangible capital, given their physical nature and the diseconomies of scale, tend to be finite. Returns on intangible assets are almost infinite. Infinite returns to scale annul the iron law of diminishing marginal returns that governed industrial society. Corporate reliance and strategy in knowledge societies are therefore largely directed toward the generation and purchase of

Intellectual Property Rights (IPRs — patents, copyrights, trademarks, brands, digital platforms). IPRs, in turn, are political creatures (Stehr, 2022; Stehr, 2023).

4. Patent law, mitigation and adaptation: The social control of knowledge (or, knowledge politics, cf. Stehr, 2005) legally formalized mainly by (international) patent laws, will assume a significant role in global efforts (cf. Young, 2021) to mitigate greenhouse gas emissions and to adapt to climate change. Whether patenting will play a crucial role in global efforts or moving relevant knowledge to the knowledge commons and making it freely available is a highly contentious issue. For example, as Hardin (2020:611; Chavez, 2015) maintains, "the number of inventions in the field of climate engineering, or 'geoengineering,' has skyrocketed over the past several years, and the number of patent applications and grants for technologies in that field has similarly increased dramatically." He views the constraining of knowledge as a favorable condition (as leverage) for the development of new climate change technologies and information. The very effort to constrain knowledge and technologies can be viewed, in contrast, as a major hurdle to the rapid and widespread dissemination of climate change-relevant knowledge (cf. Boldrin and Levine, 2013).

Some practical knowledge issues are exemplified in the following summary of ongoing research issues:

1. Climate governance for a resilient society. The dichotomy that best captures the discussion of governing a warmer climate points to approaches that favor large, multinational social organizations and large social institutions with imply top-down pledges and policies. After all, much of the narrative of climate change policies evolved during the height of the discussion on globalization (cf. Chakrabarty, 2017). The contrasting position supports much smaller social organizations and broad-based participation as effective agents of governing. If the reduction of emissions is foregrounded in climate policies, large governments and institutions are emphasized; once adaptation is stressed, irrespective of the failure or the success of national and international policy activities, individual consumers, communities, and cities become the major actors.

   Our concern is that a narrowing epistemic commitment to a technocratic approach, for instance, through a stress on algorithms, (computer-powered) models and (big) data, is driving out a democratic approach as a way of doing climate governance (cf. Edwards, 2010). As an at least implicit statement about the prospect of solving or finding future ways of coping with climate change, the technocratic approach incorporates an essentially optimistic, hopeful message about its ability to combat climate. Moreover, the technocratic approach tends to be connected to the idea that climate change is a problem for international organizations, big governments and large corporations to solve (McLeod, 2020). We are suggesting that prospects for a resilient planet require support from the bottom-up. Such a narrative needs, in any event, a reflection about conceptualizing climate change as a societal and environmental issue (also Hirsch and Long, 2021).

2. A technocratic approach misconceives the "climate problem". The climate problem is a "wicked problem" rather than a discrete problem "to be solved, climate change is better understood as a persistent condition that must be coped with and can only be partially managed more — or less — well. It is just one part of a larger complex of such conditions encompassing population, technology, wealth disparities, resource use, etc. Hence, it is not straightforwardly an 'environmental' problem either. It is axiomatically as much an energy problem, an economic development problem or a land-use problem, and may be better approached through these avenues than as a problem of managing the behavior of the Earth's climate by changing the way that humans use energy" (Prins *et al.*, 2010:16).

   What makes a problem "wicked" is the "impossibility of giving it a definitive formulation: the information needed to understand the problem is dependent upon one's idea for solving it. Furthermore, wicked problems lack a stopping rule: we cannot know whether we have a sufficient understanding to stop searching for more understanding. There is no end to causal chains in interacting open systems of which the climate is the world's prime example. So, every wicked problem can be considered as a symptom of another problem" (Prins *et al.*, 2010:16).

   Hence, work toward a resilient society requires efforts both with respect to mitigation and adaptation, as we have tried to stress early on in our essays reprinted here. Adaptive activities, depending on the prevailing ecosystems, will look different in different parts of the world. But the impact of a changing climate on everyday life becomes more noticeable in every part of the world every year; for example, increased heat exposure from a warming climate and the urban heat island effect will impact all growing urban settlements worldwide (cf. Tuholske *et al.*, 2021).

3. Fit for 55. One of the crucial global political arenas in the struggle with a warmer world is the European Union. Within Europe, climate policies are mainly made in Brussels. Will Europe become the role model for the rest of the world? The Commission of the EU has advanced its "fit for 55" program. The EU Commission proposes to reduce its CO_2 budget by 55 percent in 2030. Behind China and the United States, the EU has the third-highest volume of CO_2 emissions.

Only if the 27 states of the European Union retain their prosperity by 2030 and reach their mitigation goal can one expect that the EU will become the role model for the world. The enlarged mitigation policy will encounter its resistance, and not only from some governments. The resistance is of course fueled by the opposition to an extension of the sectors of the economy to which emission trading will be applied and to an increase in the price of CO_2, as part of the EU emission trading program. The program will move beyond the present sector targets and, in addition to energy and industrial production, include the CO_2 emissions produced by traffic and heating. In response to the higher CO_2 price, the opposition expects a severe contraction of economic growth and an increase — from country to country — in economic inequality. One driver to garner the agreement of the opposition to the "fit for 55" program of the EU are the funds generated by the emissions trading program and its clever distribution.

A further mitigation policy initiative of the EU is already in the planning stage: The "Green deal" designed to completely eliminate greenhouse gas emissions by 2050. The most contentious element of the new policy concerns a decision about the definition of nuclear energy as a sustainable form of energy generation or as non-green energy. Proponents of defining nuclear energy as green energy argue that the goal of greenhouse gas neutrality of Europe cannot be reached without incorporating the output of atomic energy plants in the calculation. A difficult policy element of the "Green deal" is the elimination of CO_2 emission produced by housing. Housing greenhouse gas emissions account for a third of all emissions of the EU. Whether the available financial means, the material resources and the human capital will be sufficient to achieve the housing goals is highly contentious.

What is to be done?

If climate change is to be addressed effectively in the medium and long run, three novel policy approaches should be considered: (1) All societies must engage in mitigation and adaptation strategies. (2) Major revenue has to be secured to cover the costs for these strategies. (3) International patent law designed more than 120 years ago (cf. Thurow, 1997) should be adjusted to the realities of climate change. Intellectual property rights regimes should not serve as vehicles limiting access to knowledge or inventions that are important tools for mitigating and adapting to climate change.

Assuming that there is growing consensus that adaptation strategies are indispensable, we have but one proposal when it comes to the question of "what is to be done" in the policy field of financing these activities. Our proposal is for the establishment of a "universal carbon revenue fund" (UCR). The volume of the investments required to reach the ambitious goals of a carbon-neutral world will have to be extraordinary. A novel way of financing the investments is required. UCR fits this bill.

During the pandemic, concern that the prevailing international system of intellectual property rights offers excessive protection to patent holders has not been voiced for the first time; but under the contingencies of vaccinating the global population speedily, the demand to keep know-how about vaccines in the public domain grew more persistent and louder. This not only favors the global patent regime patentholders; the major share of the benefits of the patents typically go to corporations in the rich countries.

In a globally integrated economy — from which developing countries and emerging markets have benefited enormously in many ways — global rules matter. The global rules have always been set to favor high-income countries; they are, to a large extent, set by the large powerful countries, and frequently by powerful special interests within them, whereas developing countries don't have a seat at the table, or are at least underrepresented (Korinek and Stiglitz, 2021:341).

A similar controversial discussion will be sparked in the coming years by the question of the extent to which the widespread patenting of technical (and other) innovations in the field of climate protection and the protection of society from climate change is justified. This fact, in turn, makes the question of legal restrictions on access to this knowledge a contentious issue in which familiar political, economic and legal positions will then clash again (see Biddle, 2016): How important is patent law for additional knowledge? Our position is exempt certain inventions from patent protection, for example, in the area of climate adaptation and life-saving intellectual property (i.e., patents designed to attend to global emergencies such as the pandemic or global warming; patents [need for knowledge] pertaining to global public goods, cf. Frow, 1996). Keep technological advances necessary to attain a carbon-zero world in the public domain.

References

Boldrin, Michele and David K. Levine (2013), "The case against patents" *Journal of Economic Perspectives* 27:3–22.

Biddle, Justin B. (2016), "Intellectual property rights and global climate change: Toward resolving an apparent dilemma" *Ethics, Policy & Environment* 19:301–319.

Block, Fred and Matthew R. Keller (2012), "Where do innovations come from? Implications for intellectual property," pp. 81–103 in Leonardo Burlamaqui, Ana Célia Castro and Rainer Kattel (eds.), *Knowledge Governance. Reasserting the Public Interest.* New York: Antham Press.

Buchholz, Wolfgang and Todd Sandler (2021), "Global public goods: A survey" *Journal of Economic Literature* 59:488–545.

Chakrabarty, Dipesh (2017), "The politics of climate change is more than the politics of capitalism" *Theory, Culture & Society* 34:25–37.

Chavez, Anthony E. (2015), "Exclusive rights to saving the planet: The patenting of geo-engineering inventions" *Northwestern Journal of Technology and Intellectual Property* 13:1–35.

Churchman, C. West (1967), "Wicked problems" *Management Science* 14:B141–B142.

David, Paul A. (1993), "Intellectual property institutions and the Panda's thumb: Patents, copyrights, and trade secrets in economic theory and history" pp. 19–65 in Mitchel B. Wallerstein, Mary Ellen Mogee and Roberta A. Schoen (eds), *Global Dimensions of Intellectual Property Rights in Science and Technology.* Washington, D.C.: National Academy Press.

David, Paul A. (2005), "Koyaanisqatsi in cyberspace. The economics of an 'out-of-balance' regime of private property rights in data and information" pp. 81–113 in Keith E. Maskus and Jerome H. Reichman (eds), *International Public Goods and Transfer of Technology under a Globalized Intellectual Property Regime*. Cambridge: Cambridge University Press.

Edwards, Paul N. (2010), A vast machine. *Computer Models, Climate Data, and the Politics of Global Warming*. Cambridge, Massachusetts: MIT Press.

Grundmann, Reiner (2016), "Climate change as a wicked social problem" *Nature Geoscience* 9:562–563.

Grundmann, Reiner (2021), "COVID and Climate: Similarities and differences" *Wiley Interdisciplinary Reviews: Climate Change* 12.6:e737.

Grundmann, Reiner and Nico Stehr (2012), *The Power of Scientific Knowledge*. From Research to Public Policy. Cambridge: Cambridge University Press.

Grundmann, Reiner and Nico Stehr (eds.) (2009), *Society. Critical Concepts in Sociology.* Four Volumes. New York: Routledge.

Hardin, Buzz (2020), "Compulsory licensing of climate engineering patents: How embracing technology- and research-sharing strategies brings us a step closer to solving climate change" *Arkansas Law Review* 3:611–629.

Hill, Alice C. (2021), *The Fight for Climate after COVID-19*. Oxford: Oxford University Press.

Hirsch, Shana Lee and Jerrold Long (2021), "Adaptive epistemologies: Conceptualizing adaptation to climate change in environmental science" *Science, Technology & Human Values* 46:298–319.

Korinek, Anton and Joseph E. Stiglitz (2021), "Artificial Intelligence, Globalization, and Strategies for Economic Development" *NBER Working Paper* No. 28453 w28453.

Latour, Bruno (2020), "Is this a dress rehearsal?" *Wordpress* March 26.

McLeod, Kathy Baughman (2020), "Building a resilient planet" *Foreign Affairs* May/June: 54–59.

Mayer, Maximilian (2012), "Chaotic climate change and security" *International Political Sociology* 6:165–185.

National Intelligence Estimate (2021), *Climate Change and International Responses Increasing Challenges to US National Security Through 2040*. Washington, D.C.: National Intelligence Council.

Peters, B. Guy (2017), "What is so wicked about wicked problems? A conceptual analysis and a research program" *Policy and Society* 36.3:385–396.

Prins, Gwythian, Isabel Galiana, Christopher Green, Reiner Grundmann, Mike Hulme, Atte Korhola, Frank Laird, Ted Nordhaus, Roger Pielke Jr., Steve Rayner, Daniel Sarewitz, Michael Shellenberger, Nico Stehr and Hiroyuki Tezuka (2010), *The Hartwell-Paper. A New Direction for Climate Policy After the Crash of 2009.* https://eprints.lse.ac.uk/27939/1/HartwellPaper_English_version.pdf (Accessed on 2 January, 2022).

Prins, Gwythian Mark Caine, Keigo Akimoto, Paulo Calmon, John Constable, Enrico Deiaco, Martin Flack, Isabel Galiana, Reiner Grundmann, Frank Laird, Elizabeth Malone, Yuhji Matsuo, Lawrence Pitt, Mikael Roman, Andrew Sleigh, Amy Sopinka, Nico Stehr, Margaret Taylor, Hiroyuki Tezuka and Masakazu Toyoda (2013), *The Vital Spark: Innovating Clean and Affordable Energy for All.* London: LSE Academic Publishing.

Rittel, Horst and Melvin Webber (1973), "Dilemmas in the general theory of planning" *Policy Sciences* 4:154–159.

Stehr, Nico (1992), *Practical Knowledge: Applying Social Science Knowledge*. London: Sage.

Stehr, Nico (1994), *Knowledge Societies*. London: Sage.

Stehr, Nico (2001), *The Fragility of Modern Societies. Knowledge and Risk in the Information Age.* London: Sage.

Stehr, Nico (2015), "Knowledge Society, History of" pp. 105–110 in James D. Wright (editor-in-chief), *International Encyclopedia of the Social & Behavioral Science*, 2nd edition, Vol. 13. Oxford: Elsevier.

Stehr, Nico (2022), *Knowledge Capitalism*. New York: Routledge.

Stehr, Nico (2023), *Knowledge Societies*. Cheltenham: Edgar Elgar.

Teece, David J. (1992), "Strategies for capturing the financial benefits from technological innovation" pp. 509–533 in Ralph Landau, Nathan Rosenberg, and David C. Mowery (eds), *Technology and the Wealth of Nations*. Stanford, California: Stanford University Press.

Thurow, Lester C. (1997), "Needed: A new system of intellectual property rights" *Harvard Business Review* September–October:95–103.

Tuholske, Cascade, Kelly Caylor, Chris Funk, Andrew Verdin, Stuart Sweeney, Kathryn Grace, Pete Peterson and Tom Evans (2021), "Global urban population exposure to extreme heat" *PNAS* 118:1–9.

Young, Oran (2021), *Grand Challenges of Planetary Governance. Global Order in Turbulent Times.* Cheltenham: Edgar Elgar.

Watts, Nick *et al.* (2019), "The 2019 report of The Lancet Countdown on health and climate change: Ensuring that the health of a child born today is not defined by a changing climate" *Lancet* 394:1836–1878.

3. Ongoing Related Work by Hans von Storch

Apart from genuinely natural science issues, the research interests of Hans von Storch stayed with the history of climate science, the perceptions of the climate science community about the role and significance of climate science in society and the challenge of building a regional climate advisory office, or climate service, at his institute.

In doing so, a new aspect of the history of climate science was explicated, namely the role of climatology in promoting *European colonialism* — both in the sense of providing an ideological foundation by applying the climatic determinism, but also in preparing the exploitation of colonies.[1]

The work concerning the *perception of the state and role of climate science by the scientific community* was mostly carried out with a former coworker of Nico Stehr, namely the Canadian sociologist Dennis Bray.[2] Four surveys[3] were designed and conducted among international populations of climate scientists, with the first in 1996 and the last in 2013, in order to study their opinions on climate change, their views on climate models and the role of science and scientists in society and for policy-making. The surveys were prepared by running a number of in-person interviews, to determine topics and relevant questions.[4] Many of the questions remained the same throughout the various surveys, in order to enable us to assess to what extent opinions and perceptions among climate scientists remained the same or have changed over time.

Data from the survey series indicate a strong increase in agreement concerning issues of manifestation and attribution of climate change while the evaluation of climate models has changed little in the past 20 years.

A number of specific issues were studied using these surveys.

---

[1] von Storch, H. and C. Gräbel, 2018: The dual role of climatology in (German) colonialism. *Academia* DOI: 10.13140/RG.2.2.23863.62880.

[2] https://hzg.academia.edu/DennisBray.

[3] 2013 report: https://www.academia.edu/5211187/A_survey_of_the_perceptions_of_climate_scientists_2013. 2008 report: https://www.academia.edu/5068617/CliSci2008_A_Survey_of_the_Perspectives_of_Climate_Scientists_Concerning_Climate_Science_and_Climate_Change. 1996+2003 report: https://www.academia.edu/3077309/The_Perspectives_of_Climate_Scientists_on_Global_Climate_Change_1996_and_2003.

[4] Bray, D. and H. von Storch, 1996: Inside science — a preliminary investigation of the case of global warming. *MPI Report* 195, 58 pp.

- An early first effort, using only the 1996 data, revealed the post-normal character of climate science.[5] Also, it became clear that contacts to policymakers were rare, while many more scientists had contacts to media.
- The 2007 survey was employed to examine the terminology concerning two key concepts in climate science, namely, *predictions* and *projections*, as used among climate scientists. The survey data suggest that the terminology used by the Intergovernmental Panel on Climate Change was not adopted, or only loosely adopted, by a significant minority of scientists.[6]
- The 2013 survey allowed an assessment whether climate scientists would adhere to Mertonian CUDOS norms of scientific conduct: The data suggested that while CUDOS remain the overall guiding moral principles, they are not fully endorsed or present in the reported conduct of climate scientists.[7]

The analysis of whether the climate science community would find itself in a post-normal stage was particularly useful[8]. As it allowed a better understanding of the functioning of climate scientists and their interplay with the *Zeitgeist*. The concept of post-normal conditions, introduced by Jerry Ravetz and Silvio Funtowicz,[9] refers to genuinely and unavoidably uncertain conditions of scientific knowledge, as stakes are high, decisions urgent and societal values are at dispute. When these conditions prevail, then there is a tendency to de-politicize policymaking and de-scientize science, or to politicize science and to scientize policymaking. Political utility gains over methodical rigor.

[5] Bray, D. and H. von Storch, 1999: Climate science. An empirical example of postnormal science. *Bull. Am. Meteorol. Soc.* 80: 439–456.

[6] Bray, D. and H. von Storch, 2009: "Prediction" or "Projection"? The nomenclature of climate science. *Sci. Commun.* 30: 534.

[7] Bray, D. and H. von Storch, 2017: The normative orientations of climate scientists. *Sci. Eng. Ethics* 23: 1351–1367.

[8] Krauss, W. and H. von Storch, 2012: Workshop report: Postnormal science: The case of climate research. *EOS Trans. Am. Geophys. Union* 93(10), 6 March 2012.

Krauss, W. M.S. Schäfer, and H. von Storch, 2012: Post-normal climate science. *Nat. and Cult.* 7: 121–132.

von Storch, 2011: Climate science, IPCC, postnormality and the crisis of trust. In: N. Roll-Hansen (ed.): *Status i klimaforskningen. Kunnskap og usikkerhet, vitenskapelige og politiske utforderinger.* Det Norske Videnskaps-Akademi, Novus forlag - Oslo, pp. 151–182.

[9] E.g., Funtowicz, S.O., and J.R. Ravetz, 1985: Three types of risk assessment: A methodological analysis. In: C. Whipple and V.T. Covello (eds): *Risk Analysis in the Private Sector.* Plenum, New York, pp. 217–231.

After the retirement of Dennis Bray, surveys were continued with other partners, dealing not only with climate issues but also with regional environmental questions.

These results entered in the set-up of a regional climate service, which attempts not only to frame results consistent with scientific assertions but also considers the knowledge base of possible stakeholders and recipients.[10]

[10] Krauss, W. and H. von Storch, 2012: Post-normal practices between regional climate services and local knowledge. *Nat. and Cult.* 7: 213–230.
von Storch, H., I. Meinke, N. Stehr, B. Ratter, W. Krauss, R.A. Pielke Jr., R. Grundmann, M. Reckermann, and R. Weisse, 2011: Regional climate services illustrated with experiences from Northern Europe. *J. Environ. Law Policy* 1: 1–15.

4. Practical Knowledge Acquired by Hans von Storch

This section is an article which was published in 2012. Not surprisingly, the reprinted article cannot be really up to date; the organization of the Institute of Coastal Research has undergone some modifications after Hans von Storch's retirement; also, the announced publication of climate change reports has been achieved in the meantime. The main concepts and ideas continue to be relevant as a foundation for linking scientific knowledge to policy-making.

4.1 Regional climate knowledge for society†

Abstract

The present misconception of climate science and its interaction with the public is addressed. While the knowledge base about the dynamics of climate and its sensitivity to elevated greenhouse gas concentrations has been greatly expanded with broad consensus in the scientific community, the communication with the public and policy makers has not led to the implementation of efficient measures to limit man-made climate change. It is suggested that a different position be adopted, namely the building of a regional climate service, which allows public and stakeholders to consider climate knowledge in the process of dealing with climate-related problems, where this is appropriate. Thus climate science should not be the avant garde of climate policy but support the political process by providing a knowledge broker service.

4.2 Climate change and the IPCC

The Intergovernmental Panel on Climate Change (IPCC) documents and assesses scientific knowledge about ongoing climate change and perspectives thereof. The range of issues covered by the IPCC is very broad and the degree of confidence that is met by the reports of the different working groups varies substantially. In particular, the report of Working Group 1, on the "science", enjoys broad acceptance, with a number of key assertions, namely strong consensual evidence that the climate system is warming, most of this warming cannot be explained without the increase in GHG concentrations — with the present knowledge, therefore, because

†This section originally appeared von Storch, H., 2012: Regional climate knowledge for society. In: M. Trögeler and S. Lingner (eds.): *Remote Sensing and Regional Climate Change*, ESPI report 41, pp. 13–18.

of the ongoing human emissions of greenhouse gases (GHG) in the foreseeable future, the warming of the climate system will continue many decades into the foreseeable future.

The strength of agreement among climate scientists to both the fact that there is global warming ("manifestation") and that its explanation needs the effect of elevated greenhouse gas concentrations ("attribution"), has been determined over the years in a series of surveys, which have been summarized by Bray.[11] While back in 1996, manifestation was accepted by some 62% of all respondents, and attribution only by 38%, both numbers have risen to well above 90% in 2010. Thus, acceptance that warming and greenhouse gases are the major cause is almost universal among climate scientists.

Unfortunately, the IPCC failed to be explicit in documenting, for instance in its "Summary for Policy Makers," consensus on questions *lacking consensus*, such as the fate of ice sheets, sea-level projections, present change of hurricanes, present change in different types of extremes. The other two working groups have achieved less scientific authority. The unfortunate and badly managed errors in the AR4 Report of Working Group II, on impacts, as well as the failure of the Chair of Working Group III to rebuke claims of manipulation, have led to less respect among scientists for the work of these two working groups.[12]

4.3 Deciding on climate policy

Many, in particular among physical climate scientists, apply the "linear model," according to which knowledge about climate dynamics, in particular the link between greenhouse gas concentration and warming, sea level and other significant state variables, can be translated directly into a set of needed policy and market instruments. This set would minimize the sum of adaptation and abatement costs.[13] Indeed, in the public discourse, the impression is raised that after the

---

[11] Bray, D., 2010: The scientific consensus of climate change revisited. *Environ. Sci. Policy* 13: 340–350.

[12] von Storch, H., 2011: Climate science, IPCC, postnormality and the crisis of trust. In: N. Roll-Hansen (ed.): *Status i klimaforskningen. Kunnskap og usikkerhet, vitenskapelige og politiske utforderinger*. Det Norske Videnskaps-Akademi, Novus forlag — Oslo, pp. 151–182. Still No Reaction to Richard Tol's Assertion About Incorrect Statements by Edenhofer in ZDF. http://klimazwiebel.blogspot.com/2010/10/still-no-reaction-to-richard-tols.html.

[13] Hasselmann, K., 1990: How well can we predict the climate crisis? In: H. Siebert (ed.): *Environmental Scarcity — the International Dimension*. JCB Mohr, Tübingen, pp. 165–183. Nordhaus, W.D., 1991: To slow or not to slow: The economy of the Greenhouse effect. *Econ. J.* 101: 920–937.

unequivocal findings of the IPCC — as given above — a mandatory political course would be clear, namely a reduction of greenhouse gas emissions as much as possible, so that temperature increase would peak at 2 degrees or less, and then stabilize.

But, in spite of a massive public campaign based on — what is called, at least in the West: a scientific consensus and conclusion, concrete and efficient manifestations of such policy remain rare and unconvincing. Obviously, the linear model does not work. One reason is that the world is seen as essentially one-directional, namely that decisions and thus "action" would essentially flow directly from scientific understanding. Also, it is based on a rather idealized understanding of the interaction between science and the public; one idealization is that on the side of the knowledge-providers there are no conflicts about what the "facts"; science as a knowledge-broker appears monolithic.

In my understanding, the political process does not make use of scientific "truth" — whatever that may be — but on perceptions and on knowledge claims that are the result of a metamorphosis of scientific knowledge. The issue has become an issue of competing knowledge claims, which are by themselves subordinate to certain worldviews and sets of value preferences. Indeed, this had to be expected after climate science found itself in a post-normal situation, where *stakes are high, facts uncertain, decisions urgent and values in dispute.*[14] Interest-led utility is a significant driver in the research area in a postnormal phase, less so "normal curiosity."

4.4 Different knowledge claims

In my understanding, *climate change is a "constructed" issue*. People hardly experience "climate change." There are different classes of *constructions*.[15] One is *scientific*, i.e., an "objective" analysis of observations and interpretation by theories. The other is *cultural*, in particular, maintained and transformed by the public media.

The *scientific construction* describes a climate that is subject to the influence of greenhouse gases (GHGs), with the primary effect of higher temperatures and

---

[14] Funtowicz, S.O. and J.R. Ravetz, 1985: Three types of risk assessment: A methodological analysis. In: C. Whipple and V.T. Covello (eds.): *Risk Analysis in the Private Sector*. Plenum, New York, pp. 217–231.

[15] von Storch, H., 2009: Climate research and policy advice: Scientific and cultural constructions of knowledge. *Env. Sci. Policy* 12: 741–747.

related facets associated with higher GHG concentrations, and secondary effects related to dynamic changes related to cloudiness, circulation, etc. In this description, humankind is responsible for the elevated GHG presence and can limit the effect of man-made climate change by regulating the emissions of greenhouse gases. However, since substantial amounts of GHGs have already been released, the effect cannot be stopped within a few decades or years. Given the inertia of the climate as well as the economic system, the warming will continue for a while. A very substantial effort has to be made to limit the warming to 2 degrees over preindustrial levels, even if there is some doubt that it is possible at all. Thus, not only efforts for reducing the flux of GHGs into the atmosphere have to be explored by science, and possibly implemented by societies, but also measures for dealing with the unavoidable changes of the possibly limited man-made climate change need to be studied and tested.

In the scientific construction, adaptation to climate change and mitigation of man-made climate change are both key aspects of the climate issue.

The *cultural construction* describes a different system, namely a sinful humankind, which is mistreating nature — which eventually strikes back, in an act of global justice. Nature, or more specifically climate, strikes back with all kinds of extremes, prominent among them being storms and hurricanes but also floods and droughts; with rising sea levels, which will in the near future destroy large coastal and island territory. All this can be halted if GHG emissions are dramatically reduced; then, and only then, can the climate crisis, or catastrophe, be managed, and further adaptation measures will not be needed, at least no significant ones.

Of course, the two constructions are not separate; both influence each other — as is common in a post-normal situation.

The present failure of science to really influence policymaking constructively and effectively may be related to the following observations:

- The science-policy/public interaction is not an issue of the linear model of "knowledge speaks to power."
- The problem is not that the public is stupid or uneducated.
- Science has failed to respond to legitimate public questions and has instead asked: "Trust us, we are scientists."
- The problem is that scientific knowledge is confronted on the "explanation market" with other forms of knowledge. Scientific knowledge does not necessarily "win" this competition.
- The social process "science" is influenced by these other knowledge forms.

I would suggest that this situation should give rise to a change in thinking among scientists, namely to give up plans to persuade societies to implement specific policies, but to support the societal process of finding solutions to the "climate problem" by answering as objectively as possible questions about the consequences of different policies, and options and needs for regional and local adaptation measures. Instead of trying to "solve" political problems on the backstage of scientific debates, science should return to its role of an honest broker[16] and build a dialogue with the public, which goes under the name of *regional climate service*.[17]

4.5 Regional climate service

The concept of "climate service" emerged first in North America, with initial publications in governmental documents in the early 1980s and earlier.[18] Its mission and scope may be summarized as: "A N[ational] C[limate] S[ervice] identifies, produces, and delivers authoritative and timely information about climate variations and trends and their impacts on built and natural systems on regional, national, and global space scales. This information informs and is informed by decision-making, risk management, and resource management concerns for a variety of public and private users acting on regional, national, and inter-national scales. The stakeholders (and the constituency for an NCS) include public and private individuals and organizations at federal, state, and local levels ... with sensitivity to and need for climate-related information."[19] Stakeholders on different scales take different viewpoints, with national and international actors being more interested in issues related to mitigation of man-made climate change and regional and local actors more engaged in adaptation measures.

The main elements of such a climate service are[20]:

---

[16] Pielke, Jr., R.A., 2007: *The Honest Broker: Making Sense of Science in Policy and Politics*. Cambridge University Press.

[17] von Storch, H., I. Meinke, N. Stehr, B. Ratter, W. Krauss, R.A. Pielke Jr., R. Grundmann, M. Reckermann, and R. Weisse, 2011: Regional Climate Services illustrated with experiences from Northern Europe. *Zeitschrift für Umweltpolitik & Umweltrecht* 1: 1–15.

[18] For a historical perspective, refer to Changnon, S.A., P.J. Lamb, and K.G. Hubbard, 1990: Regional Climate Centers: New institutions for climate services and climate-impact research. *Bull Am. Meteorol. Soc.* 71(4): 527–537.

[19] Miles, E.L., A.K. Snover, L.C. Whitely Binder, E.S. Sarachik, P.W. Mote, and N. Mantua, 2006: An approach to designing a National Climate Service. *Proc. Natl. Acad. Sci.* 103(52): 19616–19623.

[20] Miles *et al., op. cit.*

- Serve as a clearinghouse and technical access point to stakeholders for regionally and nationally relevant information on climate, climate impacts, and adaptation; developing comprehensive databases of information relevant to specific regional and national stakeholder needs.
- Provide education on climate impacts, vulnerabilities, and application of climate information in decision making.
- Design decision-support tools that facilitate use of climate information in stakeholders' near-term operations and long-term planning.
- Provide user access to climate and climate impact experts for technical assistance in use of climate information and to inform the climate forecast community of their information needs.
- Provide researcher, modeler, and observations experts access to users to help guide direction of research, modeling, and observation activities.
- Propose and evaluate adaptation strategies for climate variability and change.

This concept fits well into the linear model discussed above, which stipulates that knowledge about the dynamics in the Earth–Society system together with an understanding about the incurred costs for adaptation and mitigation, would "solve" the climate problem, and provide decision makers with directions on how to respond to the perspective of anthropogenic climate change rationally and cost-effectively.

As part of the Climate Service data collection, quality control and archival activities, dissemination, and guidance for using such data, scenario of climate change and impacts, and links to applied research often are listed.[21] Regional and global data sets, describing recent, ongoing and possible future climate changes and impacts are important elements enabling an efficient climate service.[22]

4.6 Our activities at the Institute of Coastal Research at the Helmholtz Zentrum Geesthacht

The Institute of Coastal Research at the Helmholtz Zentrum Geesthacht (near Hamburg, Germany) describes its mission in this way: Coastal systems are under constant pressure from short and long-term natural influences, including erosion or sea-level rise due to climate change, and from human endeavors, for example, transportation, land use patterns, tourism, etc. As a means to identify the potential

[21] See Chagnon *et al., op. cit.*

[22] von Storch, H. and I. Meinke, 2008: Regional Climate Offices and Regional Assessment Reports needed. *Nat. Geosci.* 1(2): 78.

for change, sustainability, and adaptation, coastal research provides the tools, assessments, and scenarios for managing this vulnerable landscape. Research activities span both the natural and human dimensions of coastal dynamics, analyzing the coastal system in global and regional contexts, conducting assessments of the state and sensitivity of the coastal system to natural and human influences, and developing scenarios of future coastal options.

As such, the Institute claims to generate useful knowledge, which can be used mostly regional and local contexts for managing coasts, in particular with respect to climate change. Being confronted with the issue discussed above, special efforts were developed and implemented — with partners from the social sciences and humanities

These efforts comprise:

1. Analysis of the *cultural constructions of climate, climate change and impact*, including common exaggeration in the media.[23]
2. Determination of response options on the local and regional scale: mainly adaptation but also regional and local mitigation.[24]
3. *Dialogue* of stakeholders and climate knowledge brokers in "Klimabureaus."[25]
4. Analysis of *consensus* on relevant issues (climate consensus reports).[26]
5. Description of recent and present *changes* as well as projection of *possible future* changes, which are dynamically consistent and possible ("scenarios") ("CoastDat").[27]
6. Direct exchange and discussion about climate science and climate policy with individuals via a weblog.[28]

[23] E.g., Neverla, I. and H. von Storch, 2010: Wer den Hype braucht. *Die Presse*, 24 July 2010.

[24] E.g., von Storch, H., M. Claussen, and KlimaCampus Autoren Team (eds.), 2010: *Klimabericht für die Metropolregion Hamburg*. Springer Verlag, Heidelberg, Dordrecht, London, New York, 321 p.

[25] Meinke, I. and H. von Storch, 2008: Regional Cimate Offices as link between climate research and decision makers. *Extended Abstract for International Desaster Reduction Conference (IDRC)*, Davos, Switzerland, 25–29 August 2008, pp. 938–941.

Schipper, J.W., I. Meinke, S. Zacharias, R. Treffeisen, Ch. Kottmeier, H. von Storch, and P. Lemke, 2009: Regionale Helmholtz Klimabüros bilden bundesweites Netz. *DMG Nachrichten* 1: 10–12.

[26] BACC author team, 2008: *Assessment of Climate Change in the Baltic Sea Basin*. Springer Verlag, Berlin-Heidelberg, 473 pp. and von Storch, H., M. Claussen, and KlimaCampus Autoren Team (eds.), *op. cit.*

[27] Weisse, R., H. von Storch, U. Callies, A. Chrastansky, F. Feser, I. Grabemann, H. Günther, A. Plüss, T. Stoye, J. Tellkamp, J. Winterfeldt, and K. Woth, 2009: Regional meteo-marine reanalyses and climate change projections: Results for Northern Europe and potentials for coastal and offshore applications. *Bull. Am.. Meteorol. Soc.* 90: 849–860.

[28] The weblog was http://klimazwiebel.blogspot.com/. It has been discontinued in the meantime.

4.6.1 *North German Climate Office*

The North German Climate Office was set up in 2006 as an institution that enables communication between science and stakeholders,[29] i.e.: making sure that:

1. science understands the questions and concerns of a variety of stakeholders
2. stakeholders understand the scientific assessments and their limits.

The office deals specifically with issues that are covered scientifically by the home institute, i.e., various aspects dealing with climate change and climate impact in the German coastal regions. As such, typical stakeholders entail representatives and stakeholders in coastal defense, agriculture, offshore activities (energy), tourism, water management, fisheries and urban planning.

A special product is the North German Climate Atlas, which is available in German language, to meet customers' demands.[30] This web-based atlas describes possible climatic futures, as given by — so far — 12 regional climate projections, for different regions in Northern Germany (plus a region straddling the Polish/German border). Scenarios are described by an ensemble means, but also by minimum and maximum changes in the set of scenarios.

4.6.2 *Regional climate consensus reports*

Scientifically legitimate knowledge about climate, climate change and climate impacts are screened in an IPCC-like process. All literature, not only in English, is considered as long as it is published in regular scientific journals or by reputable scientific institutions (such as weather services). In a series of chapters, with responsible lead authors, issues like past and ongoing regional change, possible future change, and climate-related changes in terrestrial and marine ecosystems are covered. Prior to publication, the reports are anonymously reviewed and presented to the regional scientific public. Political or management recommendations are not made, but scientifically contested areas are emphasized. The reports are conveyed to political bodies, which use them as a basis for further deliberations.

So far, two such reports have been completed.

1. The Climate Change Assessment: Report for the Baltic Sea Catchment — BACC. Approximately 80 scientists from 10 countries documented and

---

[29] http://www.norddeutsches-klimabuero.de and Meinke , I. and H. von Storch, *op. cit.*

[30] http://www.norddeutscher-klimaatlas.de/.

assessed the published knowledge, which was published in English in 2008.[31] The assessment has been employed by the intergovernmental Helsinki Commission/Baltic Marine Protection Commission (HELCOM) for the Baltic Sea as a basis for its future deliberations.[32] For 2013, the publication of an updated assessment report (BACC II) is presently being prepared.[33]

Climate Assessment for the Metropolitan Region of Hamburg. In 2007–2010, a climate assessment report about the scientifically documented knowledge of climate change in the region of Hamburg was prepared — as an activity of the Climate Centre of Excellence CLISAP at the University of Hamburg, jointly operated with the *Helmholtz Zentrum Geesthacht* and the *Max Planck Institute of Meteorology*.

2. The Senate of Hamburg and the Environmental Ministry of Schleswig Holstein used the results for climate adaptation planning.

4.6.3 *CoastDat. Regional and local conditions in the recent past and next century*

Using a modeling strategy that processes homogeneous multi-decadal analyses of large-scale circulation with a regional climate model (dynamical downscaling), a realistic description of the weather stream since 1948 until (almost) today is constructed. This description is not error-free, but the statistics of these errors remain uniform throughout the entire time. In a similar way, scenarios of possible future conditions are generated.

The whole data set, which covers atmospheric and oceanographic data, is named CoastDat.[34] It features long (60 years) and high-resolution reconstructions of recent offshore and coastal conditions mainly in terms of wind, storms, waves, surges and currents and other variables in Northern Europe, and scenarios (100 years) of possible consistent futures of coastal and offshore conditions. Efforts are underway to extend the data set, so as to cover ecological variables, but also other regions such as the Baltic Sea, East Asia and Laptev Sea.

Users of this data are various *governmental/municipal* coastal *agencies* dealing with coastal defense and coastal traffic, *companies* with needs for the assessment of risks (ship and offshore building and operations) and opportunities (wind

[31] Reckermann, M., H.-J. Isemer, and H. von Storch, (2008): Climate change assessment for the Baltic Sea basin. *EOS Trans. Am. Geophys. Union* 2008: 161–162, and BACC author team, *op. cit.*

[32] http://www.helcom.fi/. Helsinki Commission, 2007: Climate change in the Baltic Sea area. HELCOM thematic assessment in 2007. *Baltic Sea Environment Proceedings* 111.

[33] http://www.baltex-research.eu/organisation/bwg_bacc2.html.

[34] https://www.coastdat.de/, and Weisse *et al.*, *op. cit.*

energy) and finally the *general public/media*, who ask for explanations of causes of change and perspectives and options on how to deal with change.

The CoastDat-effort is pursued in cooperation with a variety of governmental agencies and also with companies. Applications cover isues such as ship design, navigational safety, assessment of offshore wind potentials, interpretations of measurements, assessments of oil spill risks and chronic oil pollution, assessment of ocean energy perspectives as well as scenarios of possible future surge and wave conditions.

4.7 Concluding remarks

When discussing the issue "knowledge for society," one has to determine what the task of science should, or could be, when interacting with society. My perspective is that this task is to:

1. offer explanations for a complex world, its dynamics, links and dependencies.
2. state what can be done, not what needs to be done.
3. establish measures to ensure the quality of science by insisting on scientific method (cf. Merton's CUDOS).
4. keep in mind that the capital of science is not the utility of the scientific findings but the methodology used to obtain such findings.

Merton CUDOS norms are repeated here; certainly, no strict rules, but a guidance, and with question marks as to what extent these rules are actually applied by wide segments of science.[35]

1. **C**ommunalism: The common ownership of scientific discoveries, according to which scientists give up intellectual property rights in exchange for recognition and esteem.
2. **U**niversalism: According to which claims to truth are evaluated in terms of universal or impersonal criteria, and not on the basis of race, class, gender, religion, or nationality.
3. **D**isinterestedness: Scientists, when presenting their work publicly, should do so without any prejudice or personal values and do so in an impersonal manner.

[35] Merton, R.K., 1978: The normative structure of science. In: N. W. Storer (ed.): *The Sociology of Science*. University of Chicago Press, Chicago, IL, pp. 267–273. Stehr, N., 1978: The ethos of science revisited social and cognitive norms. *Sociol. Inq.* 48: 172–196.

4. **O**rganized skepticism: All ideas must be tested and are subject to rigorous, structured community (peer review) scrutiny.[36]

I suggest using these rules in particular in climate sciences, as this may be a way to leave the swirl of post-normal sciences and help to lead climate science back to normal conditions. In the present situation, the policy-making process points to science when decisions are needed, even if there are difficult, value-based problems (*scientizing policymaking*). Science cannot solve these problems. But when it tries it sells out the capital of science, namely the trust of the public that science will deliver in the spirit of Merton's rules. On the other hand, if science openly takes value-based positions in favor of one or other political agenda (*politicizing science*), the foundations of good science will be destroyed.

My take-home messages for the reader are:

1. The societal service of science is to provide explanation of complex phenomena, using the scientific methodology as per Merton (CUDOS).
2. Climate science operates in a post-normal situation, which goes along with a tendency of politicizing science and scientizing politics. Cultural science needs to support climate science to deal with this challenge.
3. Climate science needs to offer "Climate Service," which includes the establishment of a dialogue with the public (direct or via media) and stakeholders — *recognizing the socio-cultural dynamics of the issue.*

[36] Grundmann, R., personal communication.

Index